Metal Phosphonates and Phosphinates

Metal Phosphonates and Phosphinates

Special Issue Editors
Marco Taddei
Ferdinando Costantino

MDPI • Basel • Beijing • Wuhan • Barcelona • Belgrade

Special Issue Editors

Marco Taddei
Swansea University
UK

Ferdinando Costantino
University of Perugia
Italy

Editorial Office
MDPI
St. Alban-Anlage 66
4052 Basel, Switzerland

This is a reprint of articles from the Special Issue published online in the open access journal *Crystals* (ISSN 2073-4352) from 2018 to 2019 (available at: https://www.mdpi.com/journal/crystals/special_issues/phosphonate).

For citation purposes, cite each article independently as indicated on the article page online and as indicated below:

LastName, A.A.; LastName, B.B.; LastName, C.C. Article Title. *Journal Name* **Year**, *Article Number*, Page Range.

ISBN 978-3-03928-002-5 (Pbk)
ISBN 978-3-03928-003-2 (PDF)

Cover image courtesy of Konstantinos D. Demadis.

© 2019 by the authors. Articles in this book are Open Access and distributed under the Creative Commons Attribution (CC BY) license, which allows users to download, copy and build upon published articles, as long as the author and publisher are properly credited, which ensures maximum dissemination and a wider impact of our publications.

The book as a whole is distributed by MDPI under the terms and conditions of the Creative Commons license CC BY NC ND.

Contents

About the Special Issue Editors vii

Preface to "Metal Phosphonates and Phosphinates" ix

Marco Taddei and Ferdinando Costantino
Metal Phosphonates and Phosphinates
Reprinted from: *Crystals* 2019, *9*, 454, doi:10.3390/cryst9090454 1

Y. Maximilian Klein, Nathalie Marinakis, Edwin C. Constable and Catherine E. Housecroft
A Phosphonic Acid Anchoring Analogue of the Sensitizer P1 for p-Type Dye-Sensitized Solar Cells
Reprinted from: *Crystals* 2018, *8*, 389, doi:10.3390/cryst8100389 4

Davood Zare, Alessandro Prescimone, Edwin C. Constable and Catherine E. Housecroft
Where Are the tpy Embraces in [Zn{4'-(EtO)$_2$OPC$_6$H$_4$tpy}$_2$][CF$_3$SO$_3$]$_2$?
Reprinted from: *Crystals* 2018, *8*, 461, doi:10.3390/cryst8120461 22

Stephen J.I. Shearan, Norbert Stock, Franziska Emmerling, Jan Demel, Paul A. Wright, Konstantinos D. Demadis, Maria Vassaki, Ferdinando Costantino, Riccardo Vivani, Sébastien Sallard, Inés Ruiz Salcedo, Aurelio Cabeza and Marco Taddei
New Directions in Metal Phosphonate and Phosphinate Chemistry
Reprinted from: *Crystals* 2019, *9*, 270, doi:10.3390/cryst9050270 33

Andrea Ienco, Giulia Tuci, Annalisa Guerri and Ferdinando Costantino
Mechanochemical Access to Elusive Metal Diphosphinate Coordination Polymer
Reprinted from: *Crystals* 2019, *9*, 283, doi:10.3390/cryst9060283 69

Konstantinos Xanthopoulos, Zafeiria Anagnostou, Sophocles Chalkiadakis, Duane Choquesillo-Lazarte, Gellert Mezei, Jan K. Zaręba, Jerzy Zoń and Konstantinos D. Demadis
Platonic Relationships in Metal Phosphonate Chemistry: Ionic Metal Phosphonates
Reprinted from: *Crystals* 2019, *9*, 301, doi:10.3390/cryst9060301 80

Jan Rohlíček, Daniel Bůžek, Petr Brázda, Libor Kobera, Jan Hynek, Jiří Brus, Kamil Lang and Jan Demel
Novel Cerium Bisphosphinate Coordination Polymer and Unconventional Metal–Organic Framework
Reprinted from: *Crystals* 2019, *9*, 303, doi:10.3390/cryst9060303 98

About the Special Issue Editors

Marco Taddei received his Ph.D. in 2011 from the University of Perugia (Italy). He stayed in Perugia until 2014, spending a period as a visiting scholar at the University of California, San Diego (USA). In 2015, he moved to the Paul Scherrer Institute (Switzerland), and in January 2017, he joined Swansea University (UK) as a Marie Curie Fellow. Marco was trained as an organic chemist, but ever since his Ph.D. days, his research has mainly focused on the synthesis and structural chemistry of metal–organic polymeric materials, such as metal–organic frameworks and metal phosphonates. In terms of applications, he is primarily interested in using these materials to capture carbon dioxide. He is the co-founder of novoMOF, a Swiss-based company producing metal–organic frameworks, for which he serves as a scientific advisor.

Ferdinando Costantino received his Ph.D. in Chemical Sciences in 2005 from the University of Perugia (Italy) under the supervision of Prof. Riccardo Vivani. He worked as a Post-Doc fellow at the University of Rennes in 2006–2007. From 2009 to 2014, he was an associate fellow of the CNR-ICCOM Institute in Florence. In 2012, he also worked as a visiting scholar at the Department of Chemistry and Biochemistry at the University of California San Diego. He is currently an associate professor of general chemistry and crystallography at the University of Perugia. He has authored more than 80 papers in international peer-reviewed journals. His research interests are related to synthesis and characterization of layered metal phosphonates and porous metal–organic frameworks for application in catalysis, energy conversion, and nanomedicine.

Preface to "Metal Phosphonates and Phosphinates"

Metal phosphonate and phosphinate chemistry has a long history, which began in the 1970s with the pioneering work independently carried out by Prof. Abraham Clearfield (Texas A&M University, USA) and Prof. Giulio Alberti (University of Perugia, Italy). In 1978, Alberti reported the synthesis of the first layered Zr phosphonate based on phenylphosphonic acid, whose crystal structure was then determined in 1993 by Clearfield. This Zr derivative is considered the archetypical structure of all metal phosphonates and disclosed a new chemistry based on the rational design of synthetic materials possessing tailor-made structures and properties due to the synergistic contribution of both the metal type and organic part of the linkers.

Phosphonic and phosphinic acids are linkers that can be synthesized by means of several, often easily accessible, strategies, thus affording a potentially huge number of building blocks. The combination of these ligands with alkaline, main group, transition, and rare-earth metals allows preparing robust and crystalline materials to be employed in a vast number of applications, such as ion exchange, gas sorption, molecular recognition, catalysis, and as support for biomedical purposes.

This Special Issue collects the latest contributions of several experts in the field who attended the First European Workshop on Metal Phosphonate Chemistry held in Swansea (UK) in September 2018. The workshop was a one-day event organized with the aim to open a forum of discussion for the most eminent scientists working in the field of phosphonate and phosphinate chemistry. The invited talks presented during the seminar covered a large number of topics, ranging from new synthetic strategies to porous compounds, catalysis, batteries, and biomedical applications.

Marco Taddei, Ferdinando Costantino
Special Issue Editors

Editorial

Metal Phosphonates and Phosphinates

Marco Taddei [1],* and Ferdinando Costantino [2],*

1. Energy Safety Research Institute, College of Engineering, Swansea University, Bay Campus Fabian Way, Swansea SA1 8EN, UK
2. Department of Chemistry Biology and Biotechnologies, University of Perugia, Via Elce di Sotto n. 8 06127 Perugia, Italy
* Correspondence: marco.taddei@Swansea.ac.uk (M.T.); ferdinando.costantino@unipg.it (F.C.)

Received: 29 August 2019; Accepted: 29 August 2019; Published: 31 August 2019

The present Special Issue entitled "Metal phosphonates and phosphinates" aims to collect recent and significant research papers on the fascinating chemistry of these two related families of coordination compounds. Phosphonic and phosphinic acids are P-containing linkers that can be synthesized by means of several, often easily accessible, strategies, thus affording a potentially huge number of building blocks. The combination of these ligands with alkaline, main group, transition and rare-earth metals allows to prepare robust and crystalline materials to be employed in a vast number of applications, such as ion-exchange, gas sorption, molecular recognition, catalysis and as support for biomedical purposes [1]. Metal phosphonate and phosphinate chemistry has a long history, begun in the early 1970s with the pioneering work independently carried out by Prof. Abraham Clearfield (Texas A&M University, USA) and Prof. Giulio Alberti (University of Perugia, Italy). In 1978, Alberti reported the synthesis of the first layered Zr phosphonate based on phenylphosphonic acid [2], whose crystal structure was then determined in 1993 by Clearfield [3]. This Zr derivative is considered the archetypical structure of all metal phosphonates and disclosed a new chemistry based on the rational design of synthetic materials possessing tailor-made structures and properties due to the synergistic contribution of both the metal type and organic part of the linkers. A number of dedicated reviews on this topic were published in the recent past [4–8]. However, the intensive research on new ligands of different complexity and functionality pushed this chemistry towards unexpected horizons and exciting achievements in the field of new materials discovery.

This Special Issue collects the latest contributions of several experts in the field who attended the First European Workshop on Metal Phosphonate Chemistry held in Swansea (UK) in September 2018. The workshop was a one-day event organized with the aim to open a forum of discussion for the most eminent scientists working in the field of phosphonates and phosphinates chemistry. The invited talks presented during the seminar covered a large number of topics, ranging from new synthetic strategies, porous compounds, catalysis, batteries and biomedical applications. An exhaustive overview of the workshop is collected into the collective perspective article entitled "New Directions in Metal Phosphonate and Phosphinate Chemistry" [9]. This perspective summarized all the talks given by the authors during the workshop and identified promising new avenues for research in the field. The success of the First European Workshop on Metal Phosphonate Chemistry led to the organisation of a second edition, to be held at the Federal Institute for Materials Research and Testing (BAM), in Berlin (Germany), on 24 September, 2019.

The Special Issue also presents five other contributions as original research papers. Two of these papers come from the group of Housecroft and Constable at the University of Basel and focus on phosphonate species as anchoring groups for dyes in dye-sensitized solar cells (DSCs). The one entitled "A Phosphonic Acid Anchoring Analogue of the Sensitizer P1 for p-Type Dye-Sensitized Solar Cells" [10] reports the synthesis of the first organic dye bearing a phosphonic acid anchoring group for p-type DSCs based on FTO/NiO electrodes. The performance of this dye compares well with that of the benchmark analogue containing a carboxylic acid anchoring group, suggesting that phosphonic acid dyes could be successfully employed in p-type DSCs. The second article is entitled "Where Are the tpy Embraces in [Zn{4'-(EtO)$_2$OPC$_6$H$_4$tpy}$_2$][CF$_3$SO$_3$]$_2$?" [11] and reports on the synthesis and structural characterization of homoleptic Zn complexes containing terpyridine ligands functionalised with either phosphonate or bromo groups. Structural analysis reveals that the presence of bulky diethylphosphonate pending groups introduces steric hindrance and affects the packing interactions, resulting in significant distortion of the ligand backbone, which is not observed in the bromo derivative. Another contribution to this Special Issue, entitled "Platonic Relationships in Metal Phosphonate Chemistry: Ionic Metal Phosphonates" [12] comes from the group led by Kostas Demadis at the University of Crete. In this article, a series of ionic compounds where the phosphonate species act as counteranions for positively charged metal-aquo complexes is presented. The lack of coordination between the phosphonates and the metal ions is attributed to the low nucleophilicity of oxygen atoms, due to the presence of electron-withdrawing groups that reduce the negative partial charge, as suggested by density-functional theory calculations. Finally, two articles report the synthesis of new metal phosphinates. In the paper entitled "Mechanochemical access to an elusive phosphinate-based coordination polymer" [13], Ienco and co-workers at the Italian National Research Council, University of Firenze and University of Perugia report that the water-assisted grinding of copper acetate with P,P'-ethylene diphenylphosphinic acid affords either a 2D, open framework compound with unprecedented topology, or an anhydrous analogue with a different structural arrangement, depending on the amount of water used during the synthesis (1 mL or two drops). Demel and co-workers at the Czech Academy of Sciences published a paper entitled "Novel Cerium Bisphosphinate Coordination Polymer and Unconventional Metal–Organic Framework" [14], where they report the synthesis of a pillared-layered cerium (III) coordination polymer based on the phenylene-1,4-bis(methylphosphinic acid) linker, named ICR-9, expanding the range of structures based on this linker and trivalent metals. ICR-9 is non-porous, but small changes in the synthetic procedure afford a less crystalline and defective solid, which displays porosity and can be considered as an unconventional metal–organic framework.

The contents of this Special Issue provide, also to non-specialist readers, an overview of the state-of-the-art and the recent progresses on phosphonate and phosphinate chemistry, with a focus on the synthesis of functional materials for various applications.

Conflicts of Interest: The authors declare no conflict of interest.

References

1. Clearfield, A.; Demadis, K. (Eds.) *Metal Phosphonate Chemistry*; Royal Society of Chemistry: Cambridge, UK, 2011; ISBN 978-1-84973-356-4.
2. Alberti, G.; Costantino, U.; Allulli, S.; Tomassini, N. Crystalline Zr(R-PO$_3$)$_2$ and Zr(R-OPO$_3$)$_2$ compounds (R = organic radical). A new class of materials having layered structure of the zirconium phosphate type. *J. Inorg. Nucl. Chem.* **1978**, *40*, 1113–1117. [CrossRef]
3. Poojary, M.D.; Hu, H.L.; Campbell, F.L.; Clearfield, A. Determination of crystal structures from limited powder data sets: Crystal structure of zirconium phenylphosphonate. *Acta Crystallogr. Sect. B* **1993**, *49*, 996–1001. [CrossRef]
4. Zhu, Y.P.; Ma, T.Y.; Liu, Y.L.; Ren, T.Z.; Yuan, Z.Y. Metal phosphonate hybrid materials: From densely layered to hierarchically nanoporous structures. *Inorg. Chem. Front.* **2014**, *1*, 360–383. [CrossRef]
5. Bao, S.S.; Zheng, L.M. Magnetic materials based on 3D metal phosphonates. *Coord. Chem. Rev.* **2016**, *319*, 63–85. [CrossRef]
6. Yücesan, G.; Zorlu, Y.; Stricker, M.; Beckmann, J. Metal-organic solids derived from arylphosphonic acids. *Coord. Chem. Rev.* **2018**, *369*, 105–122. [CrossRef]
7. Firmino, A.D.G.; Figueira, F.; Tomé, J.P.C.; Paz, F.A.A.; Rocha, J. Metal—Organic Frameworks assembled from tetraphosphonic ligands and lanthanides. *Coord. Chem. Rev.* **2018**, *355*, 133–149. [CrossRef]

8. Bao, S.S.; Shimizu, G.K.H.; Zheng, L.M. Proton conductive metal phosphonate frameworks. *Coord. Chem. Rev.* **2019**, *378*, 577–594. [CrossRef]
9. Shearan, S.J.; Stock, N.; Emmerling, F.; Demel, J.; Wright, P.A.; Demadis, K.D.; Vassaki, M.; Costantino, F.; Vivani, R.; Sallard, S.; et al. New Directions in Metal Phosphonate and Phosphinate Chemistry. *Crystals* **2019**, *9*, 270. [CrossRef]
10. Klein, Y.M.; Marinakis, N.; Constable, E.C.; Housecroft, C.E. A Phosphonic Acid Anchoring Analogue of the Sensitizer P1 for p-Type Dye-Sensitized Solar Cells. *Crystals* **2018**, *8*, 389. [CrossRef]
11. Zare, D.; Prescimone, A.; Constable, E.C.; Housecroft, C.E. Where Are the tpy Embraces in [Zn{4′-(EtO)$_2$OPC$_6$H$_4$tpy}$_2$][CF$_3$SO$_3$]$_2$? *Crystals* **2018**, *8*, 461. [CrossRef]
12. Ienco, A.; Tuci, G.; Guerri, A.; Costantino, F. Mechanochemical Access to Elusive Metal Diphosphinate Coordination Polymer. *Crystals* **2019**, *9*, 283. [CrossRef]
13. Xanthopoulos, K.; Anagnostou, Z.; Chalkiadakis, S.; Choquesillo-Lazarte, D.; Mezei, G.; Zaręba, J.K.; Zoń, J.; Demadis, K.D. Platonic Relationships in Metal Phosphonate Chemistry: Ionic Metal Phosphonates. *Crystals* **2019**, *9*, 301. [CrossRef]
14. Rohlíček, J.; Bůžek, D.; Brázda, P.; Kobera, L.; Hynek, J.; Brus, J.; Lang, K.; Demel, J. Novel Cerium Bisphosphinate Coordination Polymer and Unconventional Metal—Organic Framework. *Crystals* **2019**, *9*, 303. [CrossRef]

© 2019 by the authors. Licensee MDPI, Basel, Switzerland. This article is an open access article distributed under the terms and conditions of the Creative Commons Attribution (CC BY) license (http://creativecommons.org/licenses/by/4.0/).

Article

A Phosphonic Acid Anchoring Analogue of the Sensitizer P1 for p-Type Dye-Sensitized Solar Cells

Y. Maximilian Klein [†], Nathalie Marinakis [†], Edwin C. Constable and Catherine E. Housecroft *

Department of Chemistry, University Basel, CH-4058 Basel, Switzerland; max.klein@unibas.ch (Y.M.K.); nathalie.marinakis@unibas.ch (N.M.); edwin.constable@unibas.ch (E.C.C.)
* Correspondence: catherine.housecroft@unibas.ch; Tel.: +41-61-207-1008
† These authors contributed equally to this work.

Received: 18 September 2018; Accepted: 9 October 2018; Published: 12 October 2018

Abstract: We report the synthesis and characterization of the first example of an organic dye, **PP1**, for p-type dye-sensitized solar cells (DSCs) bearing a phosphonic acid anchoring group. **PP1** is structurally related to the benchmarking dye, **P1**, which possesses a carboxylic acid anchor. The solution absorption spectra of **PP1** and **P1** are similar (**PP1** has λ_{max} = 478 nm and ε_{max} = 62,800 dm^3 mol^{-1} cm^{-1}), as are the solid-state absorption spectra of the dyes adsorbed on FTO/NiO electrodes. p-Type DSCs with NiO as semiconductor and sensitized with **P1** or **PP1** perform comparably. For **PP1**, short-circuit current densities (J_{SC}) and open-circuit voltages (V_{OC}) for five DSCs lie between 1.11 and 1.45 mA cm^{-2}, and 119 and 143 mV, respectively, compared to ranges of 1.55–1.80 mA cm^{-2} and 117–130 mV for **P1**. Photoconversion efficiencies with **PP1** are in the range 0.054–0.069%, compared to 0.065–0.079% for **P1**. Electrochemical impedance spectroscopy, open-circuit photovoltage decay and intensity-modulated photocurrent spectroscopy have been used to compare DSCs with **P1** and **PP1** in detail.

Keywords: phosphonic acid; carboxylic acid; dye; p-type; dye-sensitized solar cell; anchor; solar energy conversion; nickel(II) oxide

1. Introduction

In n-type dye-sensitized solar cells (DSCs) [1–4], a wide variety of anchoring domains are used or have been proposed to attach the dye to the semiconductor surface [5], the most common being carboxylic and phosphonic acids. In practice, these anchors may actually function as carboxylates or phosphonates since the protonation state of the anchoring groups is not usually clearly defined and can have a significant impact of DSC performance [6]. Investigations of DSCs are dominated by studies of devices operating with an n-type semiconductor functioning as the photoanode. Despite the typically poor photoconversion efficiencies of DSCs with p-type semiconductors as the photocathode [7], research interest in the p-type interface and the development of new p-type dyes [8] is driven by the ultimate goal of tandem cells in which the performance of an n-type DSC can be further boosted by harnessing additional photoconversion events at the photocathode [9]. DSCs containing p-type photocathodes typically exhibit low fill-factors [10]. Enhancement of the performance of p-type DSCs is hampered by rapid recombination between injected holes and reduced sensitizer molecules and/or reduced electrolyte. This in turn prevents efficient dye regeneration by the electrolyte [11]. Furthermore, hole-transport resistance is typically high in p-type DSCs.

A major difference between two of the most common n-type and p-type semiconductors (TiO$_2$ and NiO) is their inherent light absorption. TiO$_2$ absorbs in the UV region and there is no interference with the light absorption of adsorbed dyes in the visible region. In contrast, NiO is typically grey or black, and as a consequence adsorbed dyes tend to be panchromatic and possess especially high extinction coefficients. Theoretical investigations are important in guiding the design of p-type dye sensitizers, including the choice of anchor and role of solvents (for example, [12–16]). One of the benchmarking p-type dyes is **P1** with a carboxylic acid anchoring group (Scheme 1) [17]. Under optimized DSC conditions and with an I_3^-/I^- redox couple, a DSC sensitized with **P1** achieves values of the short-circuit current density (J_{SC}) of 4.83 mA cm^{-2}, open-circuit voltage (V_{OC}) of 96 mV, and photoconversion efficiency (η) of 0.145% [18]. The value of J_{SC} has been increased to 7.4 mA cm^{-2} by judicious extension of the conjugated system while retaining the basic core structure of **P1** [19]. Another commonly used p-type dye is the coumarin **C343** (Scheme 2) which, like **P1**, contains a carboxylic acid anchor. Theoretical studies have indicated that stronger binding of the coumarin dye to NiO is achieved if the CO$_2$H anchor is replaced by a phosphonic acid. Moreover, calculated values of V_{OC} are influenced by a change in anchor, and the highest V_{OC} values are found for a monodentate binding mode to NiO for both CO$_2$H and PO(OH)$_2$ anchors [14]. However, to the best of our knowledge [8], there have been no experimental investigations of the use of phosphonic acid derivatives of **C343** in DSCs. Pellegrin et al. have demonstrated that the ruthenium dye **1** (Scheme 2) with values of J_{SC} = 0.78 mA cm^{-2}, V_{OC} = 95 mV and η = 0.025%, has a comparable performance to **C343** in a p-type DSC and outperforms an analogous [Ru(bpy)$_3$]$^{2+}$-based dye with CO$_2$H anchors [20]. Similarly, DSCs with the zwitterionic cyclometallated ruthenium dye **2** (Scheme 2) give photoconversion efficiencies that exceed those of cells sensitized by an analogous dye bearing a CO$_2$H anchor [21,22] and we have shown that values of J_{SC} up to 4.13 mA cm^{-2} and η up to 0.139% can be achieved using **2**, with solvent and NiO fabrication playing critical roles in optimizing the performance [23]. We were therefore motivated to investigate an analogue of dye **P1** which contains a PO(OH)$_2$ rather than CO$_2$H acid anchoring group; the structure of this **PP1** dye is shown in Scheme 1.

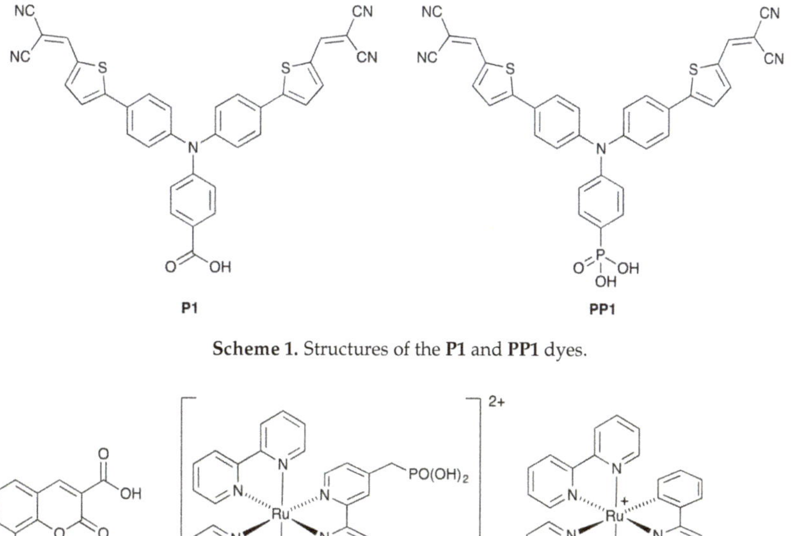

Scheme 1. Structures of the **P1** and **PP1** dyes.

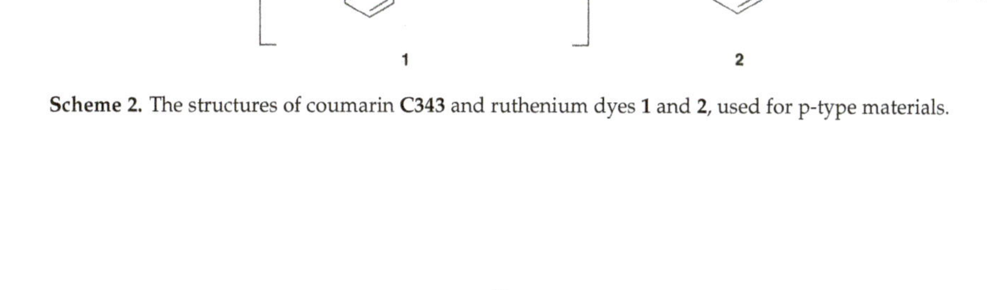

Scheme 2. The structures of coumarin **C343** and ruthenium dyes **1** and **2**, used for p-type materials.

2. Materials and Methods

2.1. General

^{1}H, ^{13}C and ^{31}P NMR spectra were recorded on a Bruker Avance III-500 spectrometer (Bruker BioSpin AG, Fällanden, Switzerland) at 295 K. The ^{1}H and ^{13}C NMR chemical shifts were referenced with respect to residual solvent peaks (δTMS = 0), and ^{31}P shifts with respect to 85% aqueous H$_3$PO$_4$. A Shimadzu LCMS-2020 instrument (Shimadzu Schweiz GmbH, Roemerstr, Switzerland) was used to record electrospray ionization (ESI) mass spectra; high resolution ESI mass spectra were recorded using a Bruker maXis 4G instrument (Bruker Daltonics GmbH, Fällanden, Switzerland) and a Bruker Daltonics Inc. microflex instrument (Bruker Daltonics GmbH, Fällanden, Switzerland) was used for MALDI mass spectra. Solution absorption spectra and solid-state absorption spectra of dye-functionalized transparent electrodes (Solaronix SA, Aubonne, Switzerland) were measured using a Cary-5000 spectrophotometer (Agilent Technologies Inc., Santa Clara, CA, USA). Reactions carried out under microwave conditions used a Biotage Initiator 8 reactor (Biotage, Uppsala, Sweden).

2.2. Compound 1

(4-Bromophenyl)diphenylamine (1.00 g, 3.08 mmol), [Pd(PPh$_3$)$_4$] (0.178 g, 0.154 mmol) and Cs$_2$CO$_3$ (1.50 g, 4.62 mmol) were added to a microwave vial and this was then evacuated and flushed with N$_2$ three times. Diethyl phosphite (0.591 mL, 0.638 g, 4.62 mmol) was dissolved in dry THF (18 mL) and N$_2$ was bubbled through the solution for 20 min. The solution was added to the microwave vial which was then sealed and heated in a microwave reactor at 120 °C for 20 min. Water (15 mL) and dichloromethane (50 mL) were added. The organic phase was extracted, dried over MgSO$_4$ and the solvent removed under reduced pressure. The crude product was purified by column chromatography (SiO$_2$, 3:2 cyclohexane/ethyl acetate, R$_f$ = 0.1). Compound 1 was isolated as a yellow oil (573 mg, 1.503 mmol, 48.8%). ^{1}H NMR (500 MHz, CDCl$_3$) δ/ppm 7.59 (dd, J = 12.7, 8.4 Hz, 2H, H^{A2}), 7.29 (t, J = 7.7 Hz, 4H, H^{B3}), 7.16–7.07 (m, 6H, H^{B2+B4}), 7.01 (dd, J = 8.6, 3.4 Hz, 2H, H^{A3}), 4.11 (m, 4H, H$^{PO(O\underline{C}H2CH3)2}$), 1.32 (t, J = 7.1 Hz, 6H, H$^{PO(OCH2\underline{C}H3)2}$). ^{13}C{^{1}H} NMR (126 MHz, CDCl$_3$) δ/ppm 151.6 (d, J_{PC} = 3.3 Hz, C^{A4}), 146.7 (C^{B1}), 133.0 (d, J_{PC} = 11.1 Hz, C^{A2}), 129.6 (C^{B3}), 125.9 (C^{B2}), 124.50 (C^{B4}), 120.3 (d, J_{PC} = 15.6 Hz, C^{A3}), 118.9 (d, J_{PC} = 195.7 Hz, C^{A1}), 62.0 (d, J_{PC} = 5.4 Hz, C$^{PO(O\underline{C}H2CH3)2}$), 16.5 (d, J_{PC} = 6.6 Hz, C$^{PO(OCH2\underline{C}H3)2}$). ^{31}P NMR (202 MHz, CDCl$_3$) δ/ppm +20.1. MALDI-MS m/z 381.31 [M]$^+$ (calc. 381.16).

2.3. Compound 2

Compound 1 (1.145 g, 3.00 mmol) was dissolved in THF, and N-bromosuccinimide (1.335 g, 7.500 mmol) was added in one portion. The solution was heated at 60 °C for 16 h. Aqueous Na$_2$CO$_3$ (20 mL, 10%) was added and the mixture was extracted with ethyl acetate (3 × 50 mL). The organic phases were combined, dried over MgSO$_4$ and the solvent removed under reduced pressure. The crude product was purified by column chromatography (SiO$_2$, 2:3 cyclohexane/ethylacetate, R$_f$ = 0.4). Compound 2 was isolated as a yellow oil (920 mg, 1.71 mmol, 56.9%). ^{1}H NMR (500 MHz, CDCl$_3$) δ/ppm 7.68–7.57 (m, 2H, H^{A2}), 7.44–7.36 (m, 4H, H^{B3}), 7.02 (dd, J = 8.6, 3.3 Hz, 2H, H^{A3}), 7.00–6.95 (m, 4H, H^{B2}), 4.15 (m, 4H, H$^{PO(O\underline{C}H2CH3)2}$), 1.33 (t, J = 7.1 Hz, 6H, H$^{PO(OCH2\underline{C}H3)2}$). ^{13}C{^{1}H} NMR (126 MHz, CDCl$_3$) δ/ppm 150.7 (d, J_{PC} = 3.3 Hz, C^{A4}), 145.6 (C^{B1}), 133.3 (d, J_{PC} = 11.0 Hz, C^{A2}), 132.9 (C^{B3}), 127.0 (C^{B2}), 121.3 (d, J_{PC} = 15.6 Hz, C^{A3}), 120.9 (d, J_{PC} = 194.9 Hz, C^{A1}), 117.5 (C^{B4}), 62.2 (d, J_{PC} = 5.5 Hz, C$^{PO(O\underline{C}H2CH3)2}$), 16.5 (d, J_{PC} = 6.5 Hz, C$^{PO(OCH2\underline{C}H3)2}$). ^{31}P NMR (202 MHz, CDCl$_3$) δ/ppm + 19.3. MALDI-MS m/z 540.13 [M + H]$^+$ (calc. 539.98).

2.4. Compound 3

5-(4,4,5,5-Tetramethyl-1,3,2-dioxaborolan-2-yl)thiophene-2-carbaldehyde (1.029 g, 4.32 mmol), Pd(PPh$_3$)$_4$ (0.125 g, 0.108 mmol) and Cs$_2$CO$_3$ (2.111 g, 6.48 mmol) were added to a microwave vial and then the vial was evacuated and flushed with N$_2$ three times. Compound 2 (0.582 g, 1.08 mmol) was dissolved in dry toluene (18 mL) and N_2 was bubbled through the solution for 20 min. The solution was added to a microwave vial, which was then sealed and heated in a microwave reactor at 120 °C for 4 h. Water (15 mL) and ethyl acetate (50 mL) were added. The organic phase was extracted, dried over $MgSO_4$ and the solvent removed under reduced pressure. The crude product was purified by column chromatography (SiO_2, ethyl acetate, R_f = 0.4). Compound 3 was isolated as a yellow oil (402 mg, 0.668 mmol, 61.9%). ^{1}H NMR (500 MHz, $CDCl_3$) δ/ppm 9.88 (s, 2H, H^{Ald}), 7.74 (d, J = 4.0 Hz, 2H, H^{C4}), 7.70 (dd, J = 12.8, 8.6 Hz, 2H, H^{A2}), 7.61 (d, J = 8.6 Hz, 4H, H^{B3}), 7.36 (d, J = 3.9 Hz, 2H, H^{C3}), 7.20–7.14 (m, 6H, $H^{B2/A3}$), 4.12 (m, 4H, $H^{PO(OCH2CH3)2}$), 1.35 (t, J = 7.1 Hz, 6H, $H^{PO(OCH2CH3)2}$). ^{13}C{^{1}H} NMR (126 MHz, $CDCl_3$) δ/ppm 182.8 (C^{Ald}), 153.6 (C^{C2}), 150.3 (d, J_{PC} = 3.4 Hz, C^{A4}), 147.5 (C^{B1}), 142.3 (C^{C5}), 137.7 (C^{C4}), 133.4 (d, J_{PC} = 11.0 Hz, C^{A2}), 127.9 (C^{B3}), 125.4 (C^{B2}), 123.85 (C^{C3}), 122.8 (d, J_{PC} = 16 Hz, C^{A3}), 122.1 (d, J_{PC} = 195 Hz, C^{A1}) 62.4 (d, J_{PC} = 5.5 Hz, $C^{PO(OCH2CH3)2}$), 16.5 (d, J_{PC} = 6.5 Hz, $C^{PO(OCH2CH3)2}$); a signal for C^{B4} was not resolved. ^{31}P NMR (202 MHz, $CDCl_3$) δ/ppm +19.0. MALDI-MS m/z 601.54 $[M + H]^+$ (calc. 602.12).

2.5. Compound 4

Compound 3 (0.349 g, 0.58 mmol) was dissolved in anhydrous MeCN (30 mL). Malononitrile (0.084 g, 1.28 mmol) and Me_3N (4 drops) were added and the mixture was heated at reflux at 85 °C for 16 h. CH_2Cl_2 (50 mL) was added and the organic phase was washed with water (3 × 30 mL), dried over $MgSO_4$ and the solvent removed under reduced pressure. The crude product was purified by column chromatography (SiO_2, ethyl acetate, R_f = 0.3). Compound 4 was isolated as a red oil (278 mg, 0.398 mmol, 68.7%). ^{1}H NMR (500 MHz, $CDCl_3$) δ/ppm 7.79 (s, 2H, H^a), 7.76–7.70 (m, 4H, $H^{A2/C4}$), 7.64–7.62 (m, 4H, C^{B3}), 7.40 (d, J = 4.1 Hz, 2H, C^{C3}), 7.21–7.15 (m, 6H, $C^{A3/B2}$), 4.15 (m, 4H, $H^{P(O)(OCH2CH3)}$), 1.35 (t, J = 7.1 Hz, 6H, $H^{P(O)(OCH2CH3)}$). ^{13}C{^{1}H} NMR (126 MHz, $CDCl_3$) δ/ppm 155.7 (C^a), 150.6 (C^{C2}), 149.9 (d, J_{PC} = 3.5 Hz, C^{A4}), 148.0 (C^{B1}), 140.4 (C^{C4}), 134.1 (C^{C5}), 133.5 (d, J_{PC} = 10.8 Hz, C^{A2}), 128.1 (C^{B3}), 125.2 (C^{B2}), 124.3 (C^{C3}), 123.6 (d, J_{PC} = 15.5 Hz, C^{A3}), 123.2 (d, J_{PC} = 194 Hz, C^{A1}), 114.4 (C^{CN}), 113.6 (C^{CN}), 76.4 (C^b), 62.3 (d, J_{PC} = 5.6 Hz, $C^{P(O)(OCH2CH3)}$), 16.5 (d, J_{PC} = 6.5 Hz, $C^{P(O)(OCH2CH3)}$); a signal for C^{B4} was not resolved. ^{31}P NMR (202 MHz, $CDCl_3$) δ/ppm +18.6. MALDI-MS m/z 697.78 $[M + H]^+$ (calc. 698.14).

2.6. PP1

Compound 4 (0.06 g, 0.086 mmol) was dissolved in anhydrous CH_2Cl_2 (30 mL). Me_3SiBr (0.227 mL, 0.263 g, 1.72 mmol) was added and the solution stirred at room temperature for 16 h. Water (20 mL) and CH_2Cl_2 (20 mL) were added and the organic phase was washed with water (3 × 30 mL), dried over $MgSO_4$ and the solvent removed under reduced pressure. The crude product was recrystallized from ethanol/cyclohexane. PP1 was isolated as a red solid (38 mg, 0.059 mmol, 68.9%). ^{1}H NMR (500 MHz, DMSO-d_6) δ/ppm 11.03 (s, 2H, $H^{P(O)(OH)2}$), 8.64 (s, 2H, H^a), 7.95 (d, J = 4.1 Hz, 2H, H^{C4}), 7.80 (d, J = 8.4 Hz, 4H, H^{B3}), 7.77 (d, J = 4.1 Hz, 2H, H^{C3}), 7.66 (dd, J_{PH} = 12.4, J_{HH} = 8.1 Hz, 2H, H^{A2}), 7.21–7.13 (m, 6H, H^{A3+B2}). ^{13}C{^{1}H} NMR (126 MHz, DMSO-d_6) δ/ppm 155.1 (C^{C2}), 153.1 (C^a), 148.2 ($C^{A1/A4}$), 143.1 (C^{C4}), 133.3 (C^{A2}), 128.5 (C^{B3}), 127.4 ($C^{A1/A4}$), 125.4 (C^{C3}), 124.6 (C^{A3+B2}), 115.2 (C^{CN}); other ^{13}C nuclei could not be resolved. ^{31}P NMR (202 MHz, DMSO-d_6) δ/ppm +25.5. ESI-MS m/z 639.99 $[M - H]^-$ (calc. 640.07). HR ESI-MS (acetone with NaOH) m/z 640.0671 $[M - H]^-$ (calc. 640.0672).

2.7. Crystallography

Single crystal data were collected on a Bruker APEX-II diffractometer (Bruker AXS GmbH, Karlsruhe, Germany); data reduction, solution and refinement used APEX2, SuperFlip and CRYSTALS respectively [24–26]. Structure analysis used Mercury v.3.6 [27,28]. Disorder of the thiophene ring containing S20 and its aldehyde group meant that this unit had to be refined isotropically.

3: $C_{32}H_{28}NO_5PS_2$, M = 601.68, yellow block, monoclinic, space group C2/c, a = 27.7428(18), b = 9.4833(6), c = 25.0824(14) Å, β = 119.127(2)°, U = 5764.5(6) Å^3, Z = 8, D_c = 1.386 Mg m^{-3}, μ(Cu-Kα) = 2.555 mm^{-1}, T = 123 K. Total 16101 reflections, 5160 unique, R_{int} = 0.027. Refinement of 4606 reflections (360 parameters) with $I > 2\sigma(I)$ converged at final $R1 = 0.0811$ ($R1$ all data = 0.0871), $wR2 = 0.2172$ ($wR2$ all data = 0.2262), gof = 0.9393. CCDC 1861694.

2.8. Electrode Preparation

Working NiO electrodes were prepared as follows. An FTO glass plate (TCO22-7, 2.2 mm thickness, sheet resistance = 7 Ω square^{-1}, Solaronix SA, Aubonne, Switzerland) was cleaned by sonicating in surfactant (2% in milliQ water), rinsed with milliQ water and EtOH and then sonicated for 10 min in acidified EtOH. The surface was sintered at 450 °C for 30 min. A pretreatment of NiO was prepared by spin-coating (3000 rpm for 1 min) onto the clean substrates of a Ni(OAc)$_2$ solution (0.5 M) containing ethanolamine (0.5 M) in methoxyethanol. After spin-coating, the plate was sintered at 500 °C for 30 min. A layer of NiO paste (Ni-Nanoxide N/SP, Solaronix SA, Aubonne, Switzerland)) was screen-printed (90 T, Serilith AG, Switzerland) onto the pretreated FTO plate, which was then placed in an EtOH chamber for 3 min to reduce surface irregularities, and dried (125 °C heating plate, 6 min). In total, two cycles of screen printing were carried out and the resultant two-layer plate was sintered by gradually heating from room temperature to 450 °C over a period of 30 min, maintained at 450 °C for 30 min, then allowed to cool over 2 h to room temperature. The NiO electrodes were soaked in an EtOH solution of Ni(OAc)$_2$ (20 mM) containing 1% ethanolamine for 30 min at 60 °C followed by EtOH rinsing and drying in air. The sintered FTO/NiO plates were then cut to form electrodes (1 × 2 cm). The thickness of the NiO layer (≈1.0–2.5 μm) was confirmed using focused ion beam (FIB) scanning electron microscopy (REM-FEI Helios NanoLab 650).

The FTO/NiO electrodes were heated at 250 °C (20 min), then cooled to 80 °C before being placed in an MeCN solution (0.3 mM) of **P1** (Dyenamo AB, Stockholm, Sweden) or an acetone solution (0.3 mM) of **PP1**. The electrodes were removed from the solutions, washed with the same solvent as used in the dye bath, then dried in an N$_2$ stream. Commercial counter electrodes (Test Cell Platinum Electrodes, Solaronix SA, Aubonne, Switzerland) were washed with EtOH, then heated at 450 °C (hot plate) for 30 min to remove volatile organic impurities. The working and counter electrodes were combined using thermoplast hot-melt sealing foil (Meltonix 1170–25 Series, 60 μm thick, Solaronix SA, Aubonne, Switzerland) by heating while pressing them together. The electrolyte of composition I$_2$ (0.1 M), LiI (1 M) in MeCN was introduced into the DSC by vacuum backfilling. The hole in the counter electrode was closed with a hot-melt sealing foil and cover glass.

2.9. Solar Cell Measurements

The solar cell measurements were made using unmasked cells with an active area of 0.237 cm^2. The DSCs were sun-soaked from the anode side for 20 min at 1 sun irradiation. The cell was then inverted and measured immediately with a LOT Quantum Design LS0811 instrument ((LOT-QuantumDesign GmbH, Darmstadt, Germany) (100 mW cm^{-2} = 1 sun at AM1.5 and 23 °C) to obtain the current density–voltage (*J–V*) curves. The instrument software was set to a p-type measurement mode (inverted configuration), with a 360 ms settling time [22] and a voltage step of 5.3 mV. The voltage was scanned from negative to positive values.

The external quantum efficiency (EQE) measurements were made using a Spe-Quest quantum efficiency instrument (Rera Systems, Nijmegen, The Netherlands) equipped with a 100 W halogen lamp (QTH) and a lambda 300 grating monochromator (L.O.T.-Oriel GmbH & Co. KG, Darmstadt, Germany). The monochromatic light was modulated to 1 Hz by using a chopper wheel (ThorLabs Inc., Newton, NJ, USA). The cell response was amplified with a large dynamic range IV converter (Melles Griot B.V., Didam, The Netherlands) and measured using a SR830 DSP Lock-In amplifier (Stanford Research Systems Inc., Sunnyvale, CA, USA).

2.10. Electrochemical Impedance Spectroscopy (EIS), Open-Circuit Photovoltage Decay (OCVD) and Intensity-Modulated Photocurrent Spectroscopy (IMPS) Measurements

EIS and IMPS measurements were carried out on a ModuLab® XM PhotoEchem photoelectrochemical measurement system (Solartron Metrology Ltd., Leicester, UK). The impedance was measured around the open-circuit potential of the cell at different light intensities (590 nm) in the frequency range 0.05 Hz to 400 kHz [29] using an amplitude of 10 mV. The impedance data were analysed using ZView® software (Scribner Associates Inc., Southern Pines, NC, USA). The IMPS measurements were performed using a small perturbation (>5%) of the steady state illumination. Voltage decay was measured on a Modulab XM electrochemical system (Solartron Metrology Ltd., Leicester, UK).

3. Results and Discussion

3.1. Synthesis and Characterization of **PP1**

The synthetic route to compound **PP1** is summarized in Scheme 3. In our hands, yields of the Pd-catalysed cross-coupling reaction introducing the phosphonic ester group are rather low and we therefore decided to carry out this transformation as the first step of the multi-step synthesis. (4-Bromophenyl)diphenylamine was treated with HPO(OEt)$_2$ in the presence of Cs$_2$CO$_3$ with [Pd(PPh$_3$)$_4$] as catalyst under microwave conditions and **1** (Scheme 3) was isolated in 48.8% yield. Treatment of **1** with NBS gave selective bromination in the 4-positions of the unsubstituted phenyl rings. A double Suzuki-Miyaura coupling of **2** (Scheme 3) using four equivalents of 5-(4,4,5,5-tetramethyl-1,3,2-dioxaborolan-2-yl)thiophene-2-carbaldehyde under microwave conditions yielded the dialdehyde intermediate **3** (see below), although the ^{1}H NMR spectrum indicated the presence of trace quantities of a second aldehyde-containing species. The electron withdrawing dicyanovinyl groups were then introduced by reaction of **3** with malononitrile (Scheme 3), and finally, the phosphonate ester groups were deprotected by treatment with Me$_3$SiBr to yield **PP1** as a red solid after recrystallization in 68.9% yield. The ^{1}H and ^{13}C NMR spectra of the intermediates in Scheme 3 and of **PP1** were fully assigned by 2D methods, and representative spectra are given in Figures S1–S4. Whereas phosphonate ester **4** is readily soluble in most common solvents, **PP1** is poorly soluble and NMR spectra were recorded in DMSO-d_6. Figure 1 shows the ^{1}H NMR spectrum of **PP1**. Complete deprotection of **4** to the acid was confirmed by the loss of the signals for the ethyl groups δ 4.15 and 1.35 ppm. The negative mode electrospray mass spectrum of **PP1** showed a base peak at m/z 639.99 corresponding to the [M − H]$^-$ ion. The solution absorption spectrum of **PP1** consists of two bands at λ_{max} = 344 and 478 nm with values of ε_{max} = 29,300 and 62,800 dm^3 mol^{-1} cm^{-1}, respectively (Figure 2). This corresponds closely to the reported MeCN solution absorption spectrum of **P1** (λ_{max} = 345 and 468 nm with ε_{max} = 58,000 dm^3 mol^{-1} cm^{-1} at 468 nm) [17].

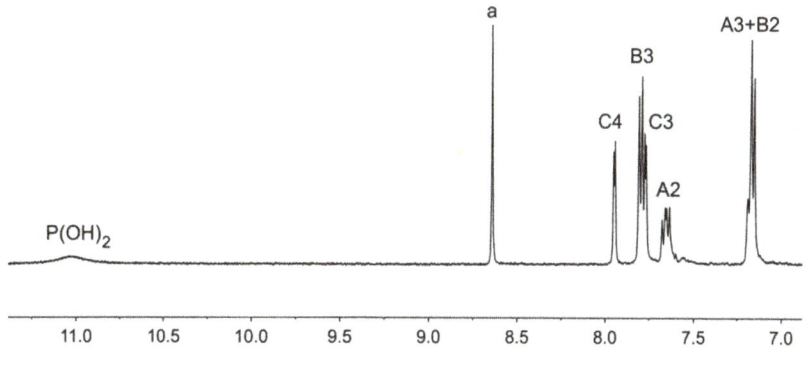

Figure 1. The ^{1}H NMR spectrum (500 MHz, DMSO-d_6) of **PP1**.

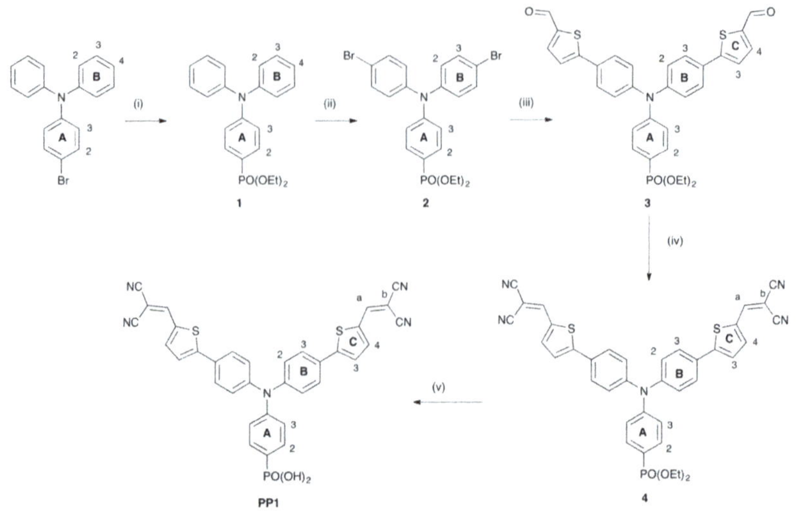

Scheme 3. Synthesis of PP1. Conditions: (i) HPO(OEt)$_2$, Cs$_2$CO$_3$, [Pd(PPh$_3$)$_4$], dry THF (120 °C, 20 min microwave reactor); (ii) NBS, THF, 60 °C; (iii) 5-(4,4,5,5-tetramethyl-1,3,2-dioxaborolan-2-yl)thiophene-2-carbaldehyde, Cs$_2$CO$_3$, [Pd(PPh$_3$)$_4$] in dry toluene (120 °C, 4 h, microwave conditions); (iv) malononitrile, Et$_3$N in dry MeCN; (v) Me$_3$SiBr, CH$_2$Cl$_2$, room temperature, 12 h, addition of H$_2$O. Ring and atom labelling are for NMR spectroscopic assignments.

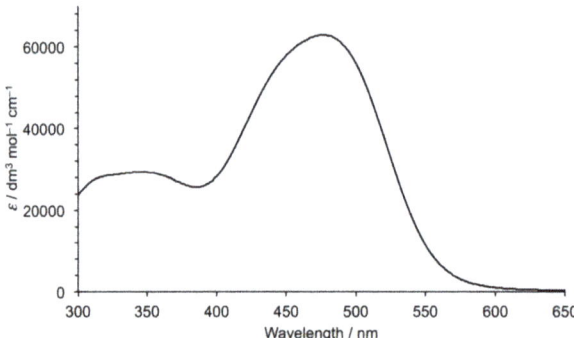

Figure 2. The solution (MeCN, 1.18×10^{-5} mol dm^{-3}) absorption spectrum of PP1.

3.2. Crystal Structure of Intermediate 3

Single crystals of the intermediate 3 were grown by slow evaporation of a solution of 3 in ethyl acetate and the structure was confirmed by single-crystal X-ray diffraction. Compound 3 crystallizes in the monoclinic C2/c space group, and the structure is depicted in Figure 3, and an ORTEP-style plot in Figure S5. One thiophene ring and its aldehyde group are disordered and was modelled over two positions with occupancies of 40 and 60%; only the major occupancy site containing S20 and C201 is shown in Figure 3 and Figure S5. The phenyl rings adopt the sterically-favored paddle-wheel arrangement around the planar N7 with N–C bond lengths and C–N–C angles in the ranges 1.403(5)–1.430(4) Å and 116.4(3)–121.7(3)°, respectively. The angles C17–C19–O20 and O1–C201–C28 of 125.5(4) and 119.9(8)° are consistent with sp^2 hybridization in the aldehyde. For the minor occupancy aldehyde, O3–C200–C30 =

118.2(8)°. The crystal structure confirms the presence of the phosphonate ester, with the bond P1–O38 being shorter than P1–O35 and P1–O39 (Figure 3 caption).

3.3. Electrode Preparation

FTO/NiO working electrodes for p-type DSCs were prepared by screen-printing two layers of NiO paste after a [Ni(OAc)$_2$] pre-treatment of the FTO surface. A pre-treatment with [Ni(OAc)$_2$] or [Ni(acac)$_2$] is essential for optimizing DSC performance by improving the adhesion of the NiO paste onto the FTO-glass surface [30,31]. In the present investigation, we modified our previously described method for Ni(acac)$_2$ pretreatment [22,23] and spin-coated a Ni(OAc)$_2$ solution in methoxyethanol containing ethanolamine onto the FTO-coated glass plate [32]. In addition, a post-treatment of Ni(OAc)$_2$ followed by sintering was applied after printing of the NiO layers in a manner similar to that described by Odobel and coworkers [32]. The final NiO surface morphology was examined using scanning electron microscopy (SEM) and fast ion bombardment (FIB) imaging (see Materials and Methods section). An NiO layer thickness of 2.0 ± 1.0 µm (Figure 4) was observed consistent with our previous work [22].

Electrodes with chemisorbed dyes **P1** or **PP1** were prepared by immersion of FTO/NiO photocathodes into solutions of the compounds. The solid-state absorption spectrum of an FTO/NiO electrode with adsorbed dye **PP1** is shown in Figure 5. The absorption band (λ_{max} = 530 nm) is red-shifted with respect to the solution spectrum (λ_{max} = 478 nm, Figure 2), consistent with the shift to lower energy reported for **P1** (468 to 499 nm) [17]. We note however, that we have also reported a value of λ_{max} = 525 nm for **P1** absorbed on FTO/NiO electrodes made in a similar manner to those used in the present work [22].

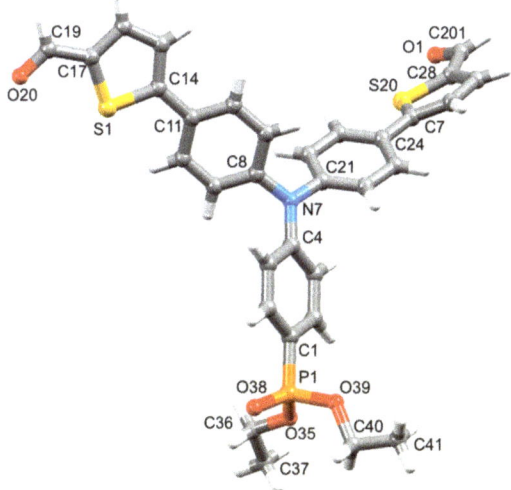

Figure 3. Single crystal structure of compound 3; an ORTEP-style plot in Figure S5. The thiophene unit containing S20 and C201 is the major occupancy site of the disordered unit (see text). Selected bond parameters: P1–O35 = 1.586(4), P1–O38 = 1.451(4), P1–O39 = 1.557(5), O20–C19 = 1.208(6), C201–O1 = 1.300(13) Å; O35–P1–O38 = 114.5(3), O35–P1–O39 = 98.7(2), O38–P1–O39 = 119.1(3), O35–P1–C1 = 107.6(2), O38–P1–C1 = 113.8(2), O39–P1–C1 = 101.2(2), C17–C19–O20 = 125.5(4), O1–C201–C28 = 119.9(8)°. For the minor occupancy aldehyde: C200–O3 = 1.238(5) Å.

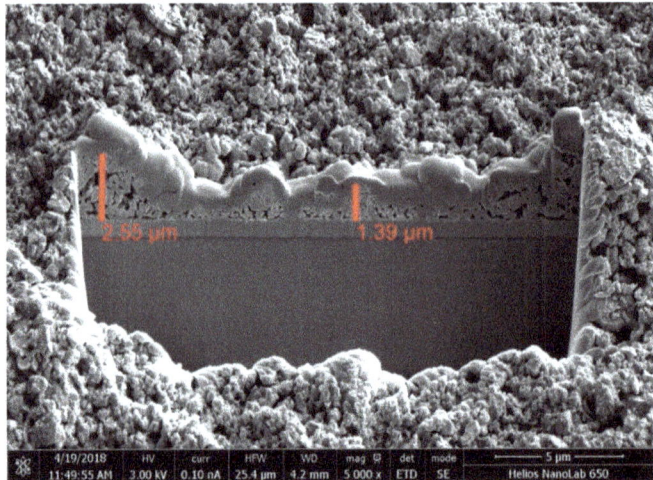

Figure 4. Fast ion bombardment (FIB) image of an NiO electrode. A platinum layer was sputtered onto the top surface as a protective layer, and a gallium beam (30 kV) was used to cut into the NiO. The depths of the NiO layer at two points (2.55 and 1.30 µm) are marked in red.

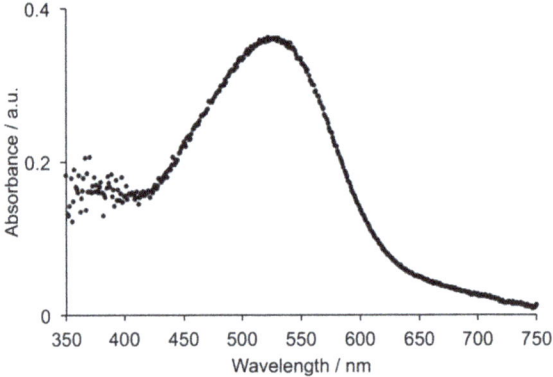

Figure 5. Solid-state absorption spectrum of FTO/NiO electrode with adsorbed dye **PP1**.

3.4. DSC Performances

Five DSCs were fabricated for each dye and the performance parameters are given in Table 1. Current-density/voltage plots are shown in Figure 6. Although the values of J_{SC} are low (not unexpected for a p-type DSC), for a given dye, the reproducibility of the cells is noteworthy. Values of J_{SC} lie in the range 1.55–1.80 mA cm^{-2} for **P1** and 1.11–1.45 mA cm^{-2} for **PP1**, and values of V_{OC} are in the range 117–130 mV for **P1** and 119–143 mV for **PP1**. The overlap of the data indicates that the dyes behave similarly, and this is borne out in the overall efficiencies (Table 1). All cells show similar fill-factors which, although low, are typical of p-type DSCs [10]. In comparison to our earlier work in which **P1** gave values of V_{OC} in the range 82–97 mV and J_{SC} values between 1.84 and 2.76 mA cm^{-2} [21–23], the DSCs in Table 1 with sensitizer **P1** achieve significantly higher values of V_{OC}, but lower values of J_{SC}. We attribute the differences to a change in the method of fabrication of the working electrodes (see Section 2.3), with the use of both pre- and post-treatments with Ni(OAc)$_2$ [32] proving beneficial.

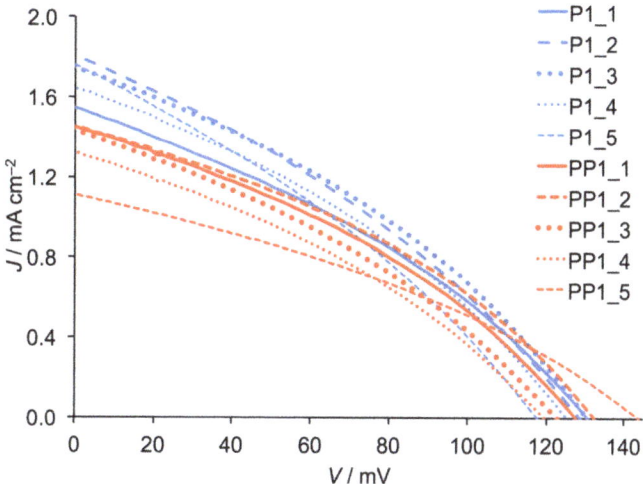

Figure 6. *J*–*V* curves for dye-sensitized solar cells (DSCs) containing dyes **P1** and **PP1**. Five replicate cells were made for each dye. Data are for the day that the cells were sealed and correspond to the values in Table 1.

Table 1. DSC performance parameters for dyes **P1** and **PP1** on NiO in p-type DSCs. Data are for the day that the cells were sealed and five DSCs were made for each dye. [a]

Dye	DSC Number	J_{SC}/mA cm^{-2}	V_{OC}/mV	*ff*/%	η/%
P1	1	1.55	130	34	0.068
P1	2	1.80	128	33	0.076
P1	3	1.74	130	35	0.079
P1	4	1.64	125	34	0.071
P1	5	1.76	117	32	0.065
PP1	1	1.45	127	35	0.065
PP1	2	1.45	132	36	0.069
PP1	3	1.32	119	34	0.054
PP1	4	1.43	122	34	0.059
PP1	5	1.11	143	34	0.054

[a] J_{SC} = short-circuit current density; V_{OC} = open-circuit voltage; *ff* = fill factor; η = photoconversion efficiency.

Figure 7 displays the external quantum efficiency (EQE) spectra for DSCs containing the two dyes. The spectra are broad (430–600 nm), consistent with charge-carrier injection over the full range of light absorption observed for the dyes (see Figure 5 and accompanying discussion). For **PP1**, EQE$_{max}$ = 10% at λ_{max}~500 nm which compares to 13.5% for **P1** (at λ_{max} = 500 nm). The latter compares to 18% reported originally by Qin et al. for **P1** [17]. We note that values of EQE$_{max}$ are significantly affected by the method of fabrication of the NiO working electrode [18], and values of up to 64% have been achieved [33]. Nonetheless, the data for **PP1** and **P1** reported here confirm the similar performances of the two dyes, indicating that the replacement of the carboxylic acid anchor in **P1** by the phosphonic acid unit in **PP1** does not have a significant detrimental effect upon electron injection.

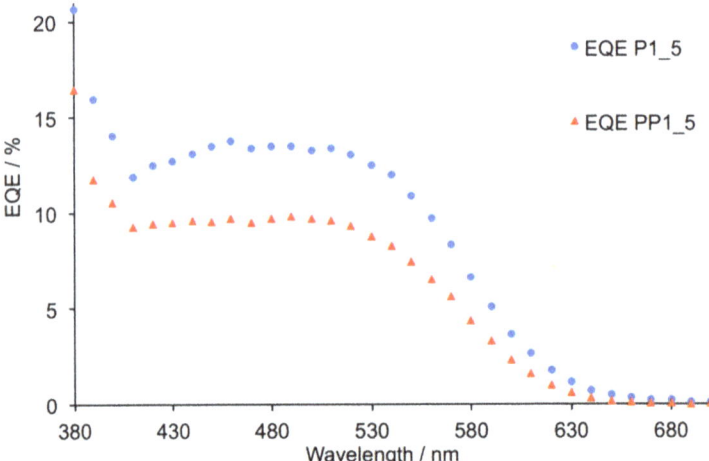

Figure 7. External quantum efficiency (EQE) spectra for DSCs with dyes **P1** and **PP1** (for each, cell 5 in Table 1).

3.5. Electrochemical Impedance Spectroscopy (EIS)

In order to compare the performances of **P1** and **PP1** in more detail, we have used EIS to investigate the internal processes and dynamics in a DSC [34–36]. A model based on an equivalent electrical circuit is used to represent the DSC. During the EIS experiment, AC voltages are applied to the DSC and the resulting current response is monitored with respect to amplitude and phase shift. Nyquist and Bode plots are used to depict the EIS results and parameters which can be extracted from the data include the recombination charge transfer resistance (R_{rec}), electron/hole transport resistance (R_t), charge-transfer resistance at the counter-electrode (R_{Pt}) and the active layer surface chemical capacitance (C_μ). A Nyquist plot comprises three semicircles, but for p-type DSCs, typically only two semicircles are seen as a consequence of the relative magnitudes of the recombination and ion diffusion impedances. Scheme 4 shows the equivalent circuit used in the present investigation, and is composed of a series resistance (R_s), a distribution element that consists of the R_t, R_{rec} and constant phase element (CPE, Q_μ) of the NiO/electrolyte interface [37]. Values of R_t, R_{rec} and Q_μ are derived from the individual values of r_t, r_{rec} and q_μ [34–36] and the resistance and constant phase element of the platinum counter electrode (R_{CE}, Q_{CE}). A CPE was used for charge transfer processes at both the NiO working electrode and platinum counter electrode due to the different porosity of the electrode surfaces [38]; we have previously detailed the equation used to calculate the fitted capacitance (C_μ) [21].

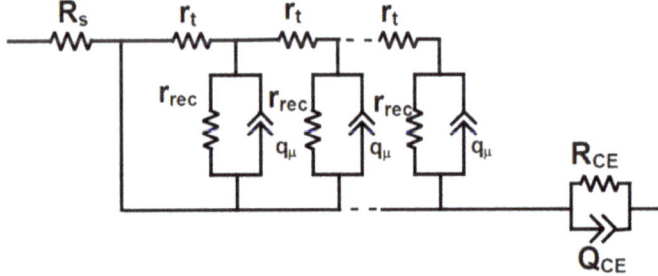

Scheme 4. Equivalent circuit used to fit the electrochemical impedance spectroscopy (EIS) data.

Figure 8 shows Nyquist plots for DSCs sensitized with **P1** and **PP1**, and parameters obtained from the fitting are presented in Table 2. The first semicircle at high frequencies (seen in the expansion in Figure 8b) is attributed to the platinum counter electrode charge transfer process, while the semicircle at lower frequencies is ascribed to charge transfer processes at the NiO/electrolyte interface. Table 2 shows that the DSC sensitized with **PP1** exhibits a higher recombination resistance (R_{rec}) while the DSC containing **P1** exhibits a lower transport resistance and a much higher capacitance. The similar overall performances of the DSCs (Table 1) indicate that these factors essentially offset one another. From the Bode plot (Figure 9), it is observed that **PP1** has the shorter hole lifetime since it has the higher frequency position [39] (f_{max} = 31.6 Hz for **PP1** versus 9.9 Hz for **P1**). Nevertheless, in order to calculate the hole lifetime of the DSC, the capacitance must first be calculated from the fitted values of the pre-factor Q and the empirical constant α (Equation (1) and Table 2) [40]. The hole lifetime can then be calculated from Equation (2). The values obtained from these equations for the two different sensitized DSCs reveal that the hole lifetime of the DSC sensitized with **PP1** is shorter than that for the device sensitized with **P1** (τ_n = 0.096 versus 0.32 ms)

$$C_\mu = \left\{ (R_{rec})^{1-\alpha} Q \right\}^{1/\alpha} \qquad (1)$$

$$\tau_n = R_{rec} C_\mu \qquad (2)$$

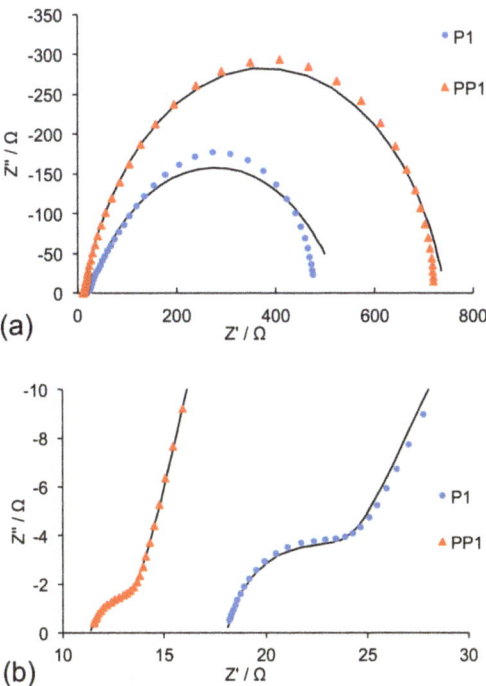

Figure 8. (a) Nyquist plots for DSCs containing **P1** or **PP1**. (b) Expansion of the high frequency region of the Nyquist plots. Fitted curves are shown as solid black lines.

Table 2. EIS data obtained from measurements at a light intensity of 22 mW cm^{-2} of p-type DSCs containing FTO/NiO working electrodes, **P1** or **PP1** dyes, and an electrolyte comprising I^-/I_3^- in MeCN.

	R_s/Ω	R_{Pt}/Ω	$C_{Pt}/\mu F$	R_{tr}/Ω	R_{rec}/Ω	$C_\mu/\mu F$	α [a]
P1	17.7	4.8	7.9	1.9	507.1	485.6	0.70
PP1	11.1	2.0	20	8.3	729.9	96.5	0.88

[a] α is an empirical constant [21].

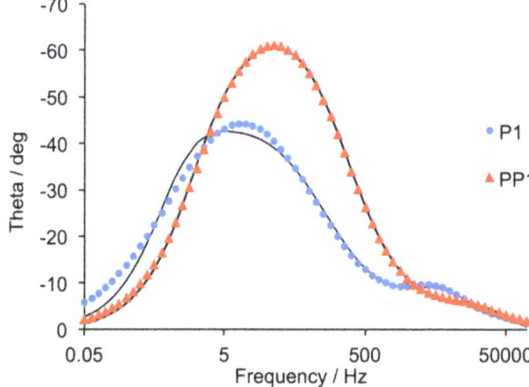

Figure 9. Bode plots for DSCs containing **P1** or **PP1**. Fitted curves are shown as solid black lines.

3.6. Open-Circuit Photovoltage Decay (OCVD) and Intensity-Modulated Photocurrent Spectroscopy (IMPS) Measurements

Under open-circuit voltage conditions, the band gap between the valence band and the conduction band of the semiconductor is at its lowest value, and the rates of charge-carrier injection and recombination are equal. From Equation (3), the lifetime of the charge, τ_n, can be extracted [41,42]. However, since the photovoltage decay is a dark measurement, the data concern only the recombination process with the electrolyte and not the recombination path with the anchored dye [41].

$$\tau_n = -\frac{k_B T}{e}\left(\frac{dV_{OC}}{dt}\right)^{-1} \tag{3}$$

Figure 10a displays the voltage decay profiles for DSCs containing dyes **P1** or **PP1**, and demonstrates that the decay of V_{OC} is more rapid for **PP1** than for **P1**. By using Equation (3), the lifetime versus voltage plot shown in Figure 10b is obtained, which shows that the lifetime of the **P1**-sensitized DSC remains longer with increased voltage. The DSC sensitized with **PP1** exhibits a greater drop in lifetime at lower voltages, which in turn translates to a more dominant charge recombination process with the electrolyte for **PP1** versus **P1**.

The diffusion length (L_d) of the charge in the NiO semiconductor is also important in defining the efficiency of the DSC and can be calculated from Equation (4) where τ_0 is the free charge lifetime and D_0 is the free charge diffusion coefficient.

$$L_d = \sqrt{\tau_0 D_0} \tag{4}$$

In an n-type DSC, the diffusion length should be 2–3 times larger than the thickness of the TiO$_2$ semiconductor for efficient charge collection [43]. However, in p-type DSCs, this ratio is difficult to achieve due to the intrinsic properties of the p-type semiconductor [9]. Intensity-modulated photocurrent spectroscopy (IMPS) [44,45] at different light intensities was used to calculate the chemical diffusion coefficient (D_n) of the **P1** and **PP1** sensitized cells. During this measurement, the cell is potentiostatically controlled and the photocurrent is measured with no bias applied. Since the measurement is done under short-circuit conditions, the semiconductor band gap is at a maximum, and no charge is exchanged at the semiconductor/electrolyte/dye interface. As a consequence, charges migrate to the back layer of the photocathode where most reactions occur, so the charge transport time, as well as the diffusion length coefficient, can be calculated [45]. From Figure 11, it can be seen that the diffusion lengths of both the **P1** and **PP1**-sensitized cells are similar and follow the same trend, only increasing slightly as the light intensity increases. This indicates a similar mechanism for the movement of the charge in the semiconductor for both dyes. The dependence of D_n on light intensity is not as great for p-type as in n-type DSCs [10], attributed to different mechanisms for the migration of charge carriers (hopping as opposed to a trapping-detrapping mechanism) [46,47].

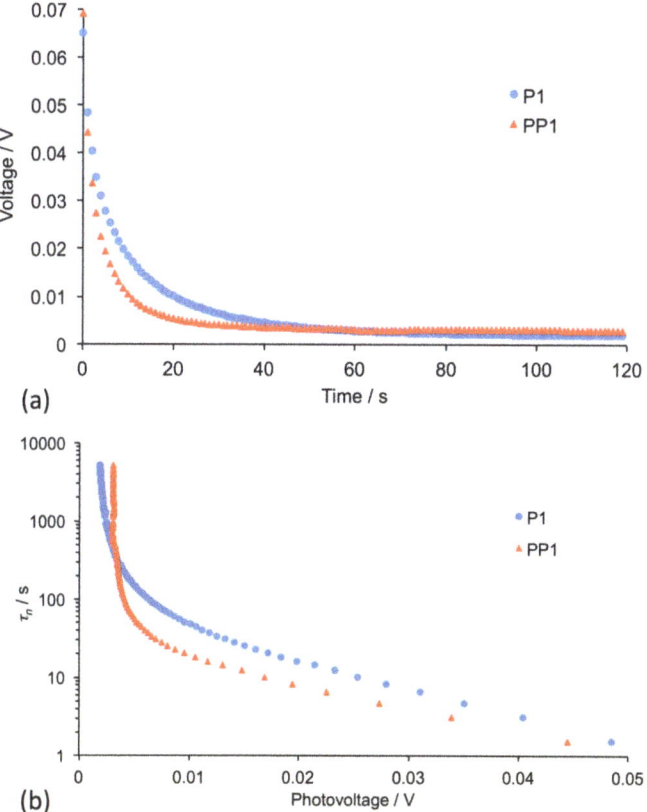

Figure 10. (a) Open-circuit voltage decay versus time. (b) Electron life time versus open circuit voltage at a light intensity of 22 mW cm^{-2}.

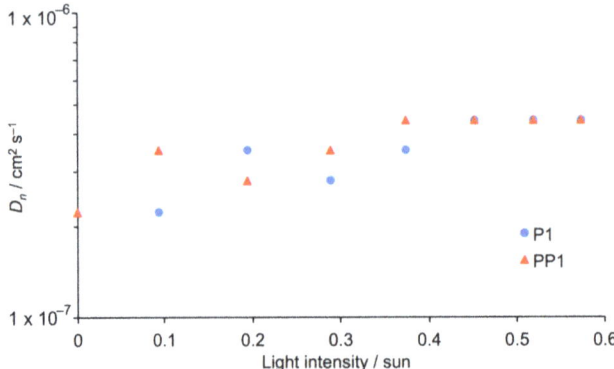

Figure 11. Chemical diffusion coefficients for DSCs with **P1** or **PP1** at different light intensities.

4. Conclusions

We report the first example of an organic dye, **PP1**, for p-type DSCs bearing a phosphonic acid anchoring group. The solution absorption spectrum of **PP1** is similar to its carboxylic acid analogue, **P1**, with the dominant absorption band at λ_{max} = 478 nm (ε_{max} = 62,800 dm^3 mol^{-1} cm^{-1}). The solid-state absorption spectrum of **PP1** adsorbed on an FTO/NiO electrode exhibits a broad band with λ_{max} = 530 nm, again similar to the spectrum for **P1**. p-Type DSCs sensitized with either **P1** or **PP1** perform comparably, confirmed using five DSCs for each dye. For **PP1**, values of J_{SC} and V_{OC} lie in the ranges 1.11–1.45 mA cm^{-2} and 119–143 mV, respectively, compared to the ranges 1.55–1.80 mA cm^{-2} and 117–130 mV for **P1**. This observation is in contrast to n-type organic dyes where those with carboxylic acid anchors typically outperform analogous dyes with phosphonic acids [5], although Abate et al. have demonstrated that in solid-state n-type DSCs, phosphonic acid anchors are beneficial [48]. We hope that the results from the present work encourage the use of phosphonic acid or phosphonate anchors in dyes for p-type DSCs.

Supplementary Materials: The following are available online at http://www.mdpi.com/2073-4352/8/10/389/s1, Figure S1: ^{1}H NMR spectrum (500 MHz, CDCl$_3$) of compound **1**, Figure S2: ^{1}H NMR spectrum (500 MHz, CDCl$_3$) of compound **2**. * = residual CHCl$_3$, Figure S3: Aromatic region of the ^{1}H NMR spectrum (500 MHz, CDCl$_3$) of compound **3**. * = residual CHCl$_3$, Figure S4: ^{1}H NMR spectrum (500 MHz, CDCl$_3$) of compound **4**. * = residual CHCl$_3$, # = H$_2$O, Figure S5: ORTEP-style plot of compound **3**.

Author Contributions: Y.M.K.: Synthesis, solution NMR and absorption spectroscopies, contributions to manuscript preparation; N.M.: all DSC measurements including EQE, EIS, OCVD, IMPS, contributions to manuscript preparation; E.C.C.: group leader, project concepts and contributions to manuscript preparation; C.E.H.: group leader, project concepts and manuscript preparation.

Funding: We thank the Swiss National Science Foundation (Grant numbers CR22I2_156236 and 200020_162631) and the University of Basel for support.

Conflicts of Interest: The authors declare no conflict of interest.

References

1. O'Reagan, B.; Grätzel, M. A low-cost, high-efficiency solar cell based on dye-sensitized colloidal TiO$_2$ films. *Nature* **1991**, *353*, 737–740. [CrossRef]
2. Nazeeruddin, Md.K.; Baranoff, E.; Grätzel, M. Dye-sensitized solar cells. A brief overview. *Sol. Energy* **2011**, *85*, 1172–1178. [CrossRef]
3. Grätzel, M. Recent Advances in Sensitized Mesoscopic Solar Cells. *Acc. Chem. Res.* **2009**, *42*, 1788–1798. [CrossRef] [PubMed]
4. Grätzel, M. Solar energy conversion by dye-sensitized photovoltaic cells. *Inorg. Chem.* **2005**, *44*, 6841–6851. [CrossRef] [PubMed]

5. Zhang, L.; Cole, J.M. Anchoring groups for dye-sensitized solar cells. *ACS Appl. Mater. Interfaces* **2015**, *7*, 3427–3455. [CrossRef] [PubMed]
6. Stephens, A.J.; Malzner, F.J.; Constable, E.C.; Housecroft, C.E. The influence of phosphonic acid protonation state on the efficiency of bis(diimine)copper(I) dye-sensitized solar cells. *Sustain. Energy Fuels* **2018**, *2*, 786–794. [CrossRef]
7. Odobel, F.; Pellegrin, Y.; Gibson, E.A.; Hagfeldt, A.; Smeigh, A.L.; Hammarström, L. Recent advances and future directions to optimize the performance of dye-sensitized solar cells. *Coord. Chem. Rev.* **2012**, *256*, 2414–2423. [CrossRef]
8. Nikolaou, V.; Charisiadis, A.; Charalambidis, G.; Coutsolelos, A.G.; Odobel, F. Recent advances and insights in dye-sensititzed NiO photocathodes for photovoltaic devices. *J. Mater. Chem. A* **2017**, *5*, 21077–21113. [CrossRef]
9. Odobel, F.; Le Pleux, L.; Pellegrin, Y.; Blart, E. New photovoltaic devices based on the sensitization of p-type semiconductors: Challenges and opportunities. *Acc. Chem. Res.* **2010**, *43*, 1063–1071. [CrossRef] [PubMed]
10. Huang, Z.; Natu, G.; Ji, Z.; He, M.; Yu, M.; Wu, Y. Probing the Low Fill Factor of NiO p-Type Dye-Sensitized Solar Cells. *J. Phys. Chem. C* **2012**, *116*, 26239–26246. [CrossRef]
11. Borgström, M.; Blart, E.; Boschloo, G.; Mukhtar, E.; Hagfeldt, A.; Hammarström, L.; Odobel, F. Sensitized Hole Injection of Phosphorus Porphyrin into NiO: Toward New Photovoltaic Devices. *J. Phys. Chem. B* **2005**, *109*, 22928–22934. [CrossRef] [PubMed]
12. Yan, W.; Chaitanya, K.; Sun, Z.-D.; Ju, X.-H. Theoretical study on p-type D-π-A snsitizers with modified π-spacers for dye-sensitized solar cells. *J. Mol. Model.* **2018**, *24*, 68. [CrossRef] [PubMed]
13. Piccinin, S.; Rocca, D.; Pastore, M. Role of solvent in the energy level alignment of dye-sensitized NiO interfaces. *J. Phys. Chem. C* **2017**, *121*, 22286–22294. [CrossRef]
14. Muñoz-García, A.B.; Pavone, M. Structure and energy level alignment at the dye-electrode interface in p-type DSSCs: New hints on the role of anchoring modes from ab initio calculations. *Phys. Chem. Phys. Chem.* **2015**, *17*, 12238–12246. [CrossRef] [PubMed]
15. Wykes, M.; Odobel, F.; Adamo, C.; Ciofini, I.; Labat, F. Anchoring groups for dyes in p-DSSC application: Insights from DFT. *J. Mol. Model.* **2016**, *22*, 289. [CrossRef] [PubMed]
16. Zhang, L.; Favereau, L.; Farre, Y.; Maufroy, A.; Pellegrin, Y.; Blart, E.; Hissler, M.; Jacquemin, D.; Odobel, F.; Hammarström, L. Molecular-structure control of electron transfer dynamics of push-pull porphyrins as sensitizers for NiO based dye sensitized solar cells. *RSC Adv.* **2016**, *6*, 77184–77194. [CrossRef]
17. Qin, P.; Zhu, H.; Edvinsson, T.; Boschloo, G.; Hagfeldt, A.; Sun, L. Design of an organic chromophore for p-type dye-sensititzed solar cells. *J. Am. Chem. Soc.* **2008**, *130*, 8570–8571. [CrossRef] [PubMed]
18. Wood, C.J.; Summers, G.H.; Clark, C.A.; Kaeffer, N.; Braeutigam, M.; Carbone, L.R.; D'Amario, L.; Fan, K.; Farre, Y.; Narbey, S.; et al. A comprehensive comparison of dye-sensitized NiO photocathodes for solar energy conversion. *Phys. Chem. Chem. Phys.* **2016**, *18*, 10727–10738. [CrossRef] [PubMed]
19. Click, K.A.; Beauchamp, D.R.; Garrett, B.R.; Huang, Z.; Hadad, C.M.; Wu, Y. A double-acceptor as a superior organic dye design for p-type DSSCs: High photocurrents and the observed light soaking effect. *Phys. Chem. Chem. Phys.* **2014**, *16*, 26103–26111. [CrossRef] [PubMed]
20. Pellegrin, Y.; Le Pleux, L.; Blart, E.; Renaud, A.; Chavillon, B.; Szuwarski, N.; Boujita, M.; Cario, L.; Jobic, S.; Jacquemin, D.; et al. Ruthenium polypyridine complexes as sensitizers in NiO based p-type dye-sensitized solar cells: Effects of the anchoring groups. *J. Photochem. Photobiol. A* **2011**, *219*, 235–242. [CrossRef]
21. Marinakis, N.; Wobill, C.; Constable, E.C.; Housecroft, C.E. Refining the anchor: Optimizing the performance of cyclometallated ruthenium(II) dyes in p-type dye sensitized solar cells. *Polyhedron* **2018**, *140*, 122–128. [CrossRef]
22. Brunner, F.; Marinakis, N.; Wobill, C.; Willgert, M.; Ertl, C.D.; Kosmalski, T.; Neuburger, M.; Bozic-Weber, B.; Glatzel, T.; Constable, E.C.; et al. Modular synthesis of simple cycloruthenated complexes with state-of-the-art performance in p-type DSCs. *J. Mater. Chem. C* **2016**, *4*, 9823–9833. [CrossRef]
23. Marinakis, N.; Willgert, M.; Constable, E.C.; Housecroft, C.E. Optimization of performance and long-term stability of p-type dye-sensitized solar cells with a cycloruthenated dye through electrolyte solvent tuning. *Sustain. Energy Fuels* **2017**, *1*, 626–635. [CrossRef]
24. Bruker X-ray Diffraction Laboratory. *M86-E01078 APEX2 User Manual*, 2nd ed.; Bruker AXS Inc.: Madison, WI, USA, 2006.

25. Palatinus, L.; Chapuis, G. SUPERFLIP. A computer program for the solution of crystal structures by charge flipping in arbitrary dimensions. *J. Appl. Cryst.* **2007**, *40*, 786–790. [CrossRef]
26. Betteridge, P.W.; Carruthers, J.R.; Cooper, R.I.; Prout, K.; Watkin, D.J. CRYSTALS Version 12: Software for Guided Crystal Structure Analysis. *J. Appl. Cryst.* **2003**, *36*, 1487–1487. [CrossRef]
27. Macrae, C.F.; Edgington, P.R.; McCabe, P.; Pidcock, E.; Shields, G.P.; Taylor, R.; Towler, M.; van de Streek, J. Mercury: Visualization and analysis of crystal structures. *J. Appl. Cryst.* **2006**, *39*, 453–457. [CrossRef]
28. Macrae, C.F.; Bruno, I.J.; Chisholm, J.A.; Edgington, P.R.; McCabe, P.; Pidcock, E.; Rodriguez-Monge, L.; Taylor, R.; van de Streek, J.; Wood, P.A. Mercury CSD 2.0—New Features for the Visualization and Investigation of Crystal Structures. *J. Appl. Cryst.* **2008**, *41*, 466–470. [CrossRef]
29. Hod, I.; Tachan, Z.; Shalom, M.; Zaban, A. Characterization and control of the electronic properties of a NiO based dye sensitized photocathode. *Phys. Chem. Chem. Phys.* **2013**, *15*, 6339–6343. [CrossRef] [PubMed]
30. Zhang, X.L.; Huang, F.; Nattestad, A.; Wang, K.; Fu, D.; Mishra, A.; Bäuerle, P.; Bach, U.; Cheng, Y.-B. Enhanced open-circuit voltage of p-type DSC with highly crystalline NiO nanoparticles. *Chem. Commun.* **2011**, *47*, 4808–4810. [CrossRef] [PubMed]
31. Perera, I.R.; Daeneke, T.; Makuta, S.; Yu, Z.; Tachibana, Y.; Mishra, A.; Bäuerle, P.; Ohlin, C.A.; Bach, U.; Spiccia, L. Application of the tris(acetlyacetonato)iron(III)/(II) redox couple in p-type dye-sensitized solar cells. *Angew. Chem. Int. Ed.* **2015**, *54*, 3758–3762. [CrossRef] [PubMed]
32. Farré, Y.; Raissi, M.; Fihey, A.; Pellegrin, Y.; Blart, E.; Jacquemin, D.; Odobel, F. A blue diketopyrrolopyrrole sensitizer with high efficiency in nickel-oxide-based dye-sensitized solar cells. *ChemSusChem* **2017**, *10*, 2618–2625. [CrossRef] [PubMed]
33. Li, L.; Gibson, E.A.; Qin, P.; Boschloo, G.; Gorlov, M.; Hagfeldt, A.; Sun, L. Double-layered NiO photocathodes for p-type DSSCs with record IPCE. *Adv. Mater.* **2010**, *22*, 1759–1762. [CrossRef] [PubMed]
34. Bisquert, J. Theory of the impedance of charge transfer via surface states in dye-sensitized solar cells. *J. Electroanal. Chem.* **2010**, *646*, 43–51. [CrossRef]
35. Fabregat-Santiago, F.; Garcia-Belmonte, G.; Mora-Seró, I.; Bisquert, J. Characterization of nanostructured hybrid and organic solar cells by impedance spectroscopy. *Phys. Chem. Chem. Phys.* **2011**, *13*, 9083–9118. [CrossRef] [PubMed]
36. Fabregat-Santiago, F.; Bisquert, J.; Palomares, E.; Otero, L.; Kuang, D.; Zakeeruddin, S.M.; Grätzel, M. Correlation between photovoltaic performance and impedance spectroscopy of dye-sensiitzed solar cells based on ionic liquids. *J. Phys. Chem. C* **2007**, *111*, 6550–6560. [CrossRef]
37. Bisquert, J.; Garcia-Belmonte, G.; Fabregat-Santiago, F.; Compte, A. Anomalous transport effects in the impedance of porous film electrodes. *Electrochem. Commun.* **1999**, *1*, 429–435. [CrossRef]
38. Córdoba-Torres, P. Relationship between constant-phase element (CPE) parameters and physical properties of films with distributed resistivity. *Electrochim. Acta* **2017**, *225*, 592–604. [CrossRef]
39. Ho, P.; Bao, L.Q.; Ahn, K.S.; Cheruku, R.; Kim, J.H. P-Type dye-sensitized solar cells: Enhanced performance with a NiO compact blocking layer. *Synth. Met.* **2016**, *217*, 314–321. [CrossRef]
40. Shoar Abouzari, M.R.; Berkemeier, F.; Schmitz, G.; Wilmer, D. On the physical interpretation of constant phase elements. *Solid State Ionics* **2009**, *180*, 922–927. [CrossRef]
41. Bisquert, J.; Zaban, A.; Greenshtein, M.; Mora-Seró, I. Determination of rate constants for charge transfer and the distribution of semiconductor and electrolyte electronic energy levels in dye-sensitized solar cells by open-circuit photovoltage decay method. *J. Am. Chem. Soc.* **2004**, *126*, 13550–13559. [CrossRef] [PubMed]
42. Zaban, A.; Greenshtein, M.; Bisquert, J. Determination of the electron lifetime in nanocrystalline dye solar cells by open-circuit voltage decay measurements. *ChemPhysChem* **2003**, *4*, 859–864. [CrossRef] [PubMed]
43. Dunn, H.K.; Westin, P.-O.; Staff, D.R.; Peter, L.M.; Walker, A.B.; Boschloo, G.; Hagfeldt, A. Determination of the electron diffusion length in dye-sensitized solar cells by substrate contact patterning. *J. Phys. Chem. C* **2011**, *115*, 13932–13937. [CrossRef]
44. Schlichthörl, G.; Huang, S.Y.; Sprague, J.; Frank, A.J. Band edge movement and recombination kinetics in dye-sensitized nanocrystalline TiO$_2$ solar cells: A study by intensity modulated photovoltage spectroscopy. *J. Phys. Chem. B* **1997**, *101*, 8141–8155. [CrossRef]
45. Dloczik, L.; Ileperuma, O.; Lauermann, I.; Peter, L.M.; Ponomarev, E.A.; Redmond, G.; Shaw, N.J.; Uhlendorf, I. Dynamic response of dye-sensitized nanocrystalline solar cells: Characterization by intensity-modulated photocurrent spectroscopy. *J. Phys. Chem. B* **1997**, *101*, 10281–10289. [CrossRef]

46. Zhu, H.; Hagfeldt, A.; Boschloo, G. Photoelectrochemistry of mesoporous NiO electrodes in iodide/triiodide electrolytes. *J. Phys. Chem. C* **2007**, *111*, 17455–17458. [CrossRef]
47. Peter, L. Transport, trapping and interfacial transfer of electrons in dye-sensitized nanocrystalline solar cells. *J. Electroanal. Chem.* **2007**, *599*, 233–240. [CrossRef]
48. Abate, A.; Pérez-Tejadam, R.; Wojciechowski, K.; Foster, J.M.; Sadhanala, A.; Steiner, U.; Snaith, H.J.; Franco, S.; Orduna, J. Phosphonic anchoring groups in organic dyes for solid-state solar cells. *Phys. Chem. Chem. Phys.* **2015**, *17*, 18780–18789. [CrossRef] [PubMed]

© 2018 by the authors. Licensee MDPI, Basel, Switzerland. This article is an open access article distributed under the terms and conditions of the Creative Commons Attribution (CC BY) license (http://creativecommons.org/licenses/by/4.0/).

Article

Where Are the tpy Embraces in [Zn{4′-(EtO)$_2$OPC$_6$H$_4$tpy}$_2$][CF$_3$SO$_3$]$_2$?

Davood Zare, Alessandro Prescimone, Edwin C. Constable and Catherine E. Housecroft *

Department of Chemistry, University Basel, BPR 1096, Mattenstrasse 24a, CH-4058 Basel, Switzerland; Davood.Zare@unibas.ch (D.Z.); alessandro.prescimone@unibas.ch (A.P.); Edwin.constable@unibas.ch (E.C.C.)
* Correspondence: catherine.housecroft@unibas.ch; Tel.: +41-61-207-1008

Received: 28 November 2018; Accepted: 7 December 2018; Published: 10 December 2018

Abstract: In this paper, the bromo- and phosphonate-ester-functionalized complexes [Zn(**1**)$_2$][CF$_3$SO$_3$]$_2$ and [Zn(**2**)$_2$][CF$_3$SO$_3$]$_2$ (**1** = 4′-(4-bromophenyl)-2,2′:6′,2″-terpyridine, **2** = diethyl (4-([2,2′:6′,2″-terpyridin]-4′-yl)phenyl)phosphonate) are reported. The complexes have been characterized by electrospray mass spectrometry, IR and absorption spectroscopies, and multinuclear NMR spectroscopy. The single-crystal structures of [Zn(**1**)$_2$][CF$_3$SO$_3$]$_2$·MeCN$^{·1}$/$_2$Et$_2$O and [Zn(**2**)$_2$][CF$_3$SO$_3$]$_2$ have been determined and they confirm {Zn(tpy)$_2$}$^{2+}$ cores (tpy = 2,2′:6′,2″-terpyridine). Ongoing from X = Br to P(O)(OEt)$_2$, the {Zn(4′-XC$_6$H$_4$tpy)$_2$}$^{2+}$ unit exhibits significant "bowing" of the backbone, which is associated with changes in packing interactions. The [Zn(**1**)$_2$]$^{2+}$ cations engage in head-to-tail 4′-Phtpy...4′-Phtpy embraces with efficient pyridine...phenylene π-stacking interactions. The [Zn(**2**)$_2$]$^{2+}$ cations pack with one of the two ligands involved in pyridine...pyridine π-stacking; steric hindrance between one C$_6$H$_4$PO(OEt)$_2$ group and an adjacent pair of π-stacked pyridine rings results in distortion of backbone of the ligand. This report is the first crystallographic determination of a salt of a homoleptic [M{4′-(RO)$_2$OPC$_6$H$_4$tpy}$_2$]$^{n+}$ cation.

Keywords: phosphonate ester; zinc(II); 2,2′:6′,2″-terpyridine; crystal structure

1. Introduction

The oligopyridines are archetypal metal-binding domains used as scaffolds for numerous functional architectures [1–3]. 2,2′:6′,2″-Terpyridines (tpy) are of particular interest as they are particularly readily accessible through simple synthetic procedures [4], and improvements to synthetic procedures continue to be reported [5]. Furthermore, implementations of oligopyridine ligands to novel target technologies continue to emerge, including optoelectronic devices [6,7]. Dye-sensitized solar cells (DSCs) utilize metal complexes as sensitizers for semiconductors [8] and the vast majority incorporate oligopyridine ligands [9]. The binding of the complex to the surface is through an anchoring ligand [10], which is typically an oligopyridine bearing carboxylic acid groups. Phosphonic acid anchoring groups have been shown to bind more strongly to semiconductor oxide surfaces than carboxylic acids and we have been particularly interested in the use of the phosphonic acid anchoring ligands shown in Scheme 1. The phosphonic acid anchor has been shown to be effective in complexes containing {Ru(bpy)$_2$(C^N)}$^+$ [11,12], {Cu(bpy)$_2$}$^+$ [13,14], and {Zn(tpy)$_2$}$^+$ [15,16] cores (C^N = a cyclometallating ligand such as the conjugate base of 2-phenylpyridine, bpy = 2,2′-bipyridine), and we have demonstrated that, in copper-based dyes, the presence of both a 1,4-phenylene spacer between a bpy metal-binding domain and the phosphonic acid is beneficial to DSC performance [17]. The tpy derivative shown in Scheme 1 has been used in a number of {Ru(tpy)$_2$}$^+$- and {Zn(tpy)$_2$}$^+$-based dyes [15–22], and theoretical studies indicate that the presence of the 1,4-phenylene spacer enhances the rate of electron-transfer across the dye/semiconductor interface [23].

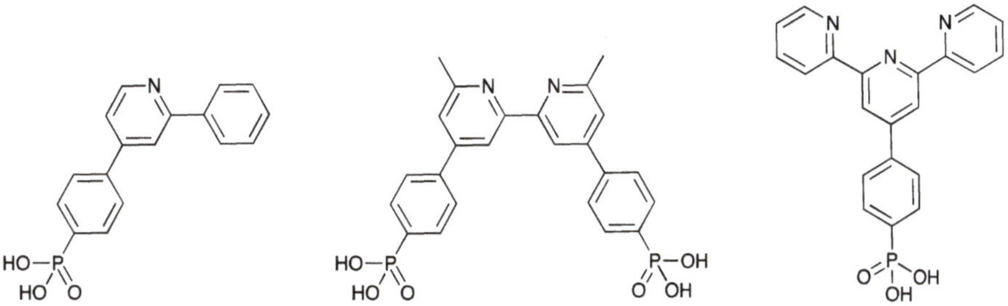

Scheme 1. Examples of anchoring ligands (cyclometallating H(C^N), bpy N^N, and tpy N^N^N) used in sensitizers in metal complex dyes in dye-sensitized solar cells.

Despite the interest in sensitizers based on polypyridine metal complexes incorporating phosphonic acid anchors, there is remarkably little crystallographic data available to provide insights into structural details, in particular, packing interactions that may influence aggregation of the surface-adsorbed species. A search of the CSD (v. 5.40, November 2018 [24] for metal-bonded {4'-O$_3$P-tpy} or {4'-O$_3$P-C$_6$H$_4$tpy} units gave surprisingly few hits [25–32], with only one featuring a 1,4-phenylene spacer [32]. It is also interesting to note that all of the reported structures relate to phosphonate esters rather than the parent phosphonic acids, and that the majority of the studies were motivated by application of the complexes as logic gates, photosensitizers, and catalysts. Here, we report the syntheses and single-crystal structures of the two zinc(II) complexes [Zn(**1**)$_2$][CF$_3$SO$_3$]$_2$ and [Zn(**2**)$_2$][CF$_3$SO$_3$]$_2$ (where ligands **1** and **2** are defined in Scheme 2) and investigate the effects that the phosphonate ester group has on intermolecular interactions in the solid state.

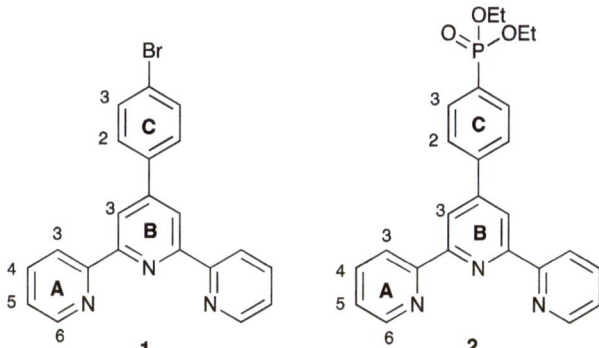

Scheme 2. Structures of the ligands **1** and **2**.

2. Materials and Methods

2.1. General

^{1}H, ^{13}C, and ^{31}P NMR spectra were recorded on a Bruker Avance III-500 spectrometer (Bruker BioSpin AG, Fällanden, Switzerland) at 298 K. The ^{1}H and ^{13}C NMR chemical shifts were referenced with respect to residual solvent peaks (δ TMS = 0), and ^{31}P shift with respect to 85% aqueous H$_3$PO$_4$. A Shimadzu LCMS-2020 instrument (Shimadzu Schweiz GmbH, Roemerstr., Switzerland) was used to record electrospray ionization (ESI) mass spectra. Solution absorption spectra were measured using a Cary 5000 spectrophotometer (Agilent, Lautengartenstr. 6, Basel, Switzerland). Solid-state IR spectra were recorded using a Perkin Elmer UATR Two spectrometer (Perkin Elmer, Bahnstrasse 8, Schwerzenbach, Switzerland).

Compound **1** was prepared by the one-pot method of Wang and Hanan [33], and compound **2** was prepared from **1** by the method reported in [34].

2.2. [Zn(**1**)$_2$][CF$_3$SO$_3$]$_2$

A solution of Zn(CF$_3$SO$_3$)$_2$ (155 mg, 0.426 mmol) in MeCN (10 mL) was added to a solution of the **1** (347 mg, 0.895 mmol) in MeCN (10 mL). The reaction mixture was heated at 55 °C for 15 h and then the product was precipitated from the colourless solution by the addition of Et$_2$O. The precipitate was collected by filtration, washed with cold CH$_2$Cl$_2$ and Et$_2$O, and dried under vacuum. [Zn(**1**)$_2$][CF$_3$SO$_3$]$_2$ was isolated as a white microcrystalline solid (437 mg, 0.383 mmol, 89.9%). Slow diffusion of Et$_2$O into a solution of the complex in MeCN gave colourless single crystals suitable for X-ray crystallography. ^{1}H NMR (CD$_3$CN, 500 MHz) δ/ppm: 8.98 (s, 4H, H^{B3}), 8.72 (dt, J = 8.1, 1.0 Hz, 4H, H^{A3}), 8.43 (td, J = 7.8, 1.6 Hz, 4H, H^{A4}), 8.13 (m, 4H, H^{C2}), 7.95 (m, 4H, H^{C3}), 7.88 (m, 4H, H^{A6}), 7.42 (m, 4H, H^{A5}). ^{13}C NMR (CD$_3$CN, 126 MHz) δ/ppm: 156.1 (C^{B4}), 150.9 (C$^{A2/B2}$), 149.0 (C^{A6}), 148.8 (C$^{A2/B2}$), 142.2 (C^{A4}), 136.3 (C^{C1}), 133.7 (C^{C3}), 130.9 (C^{C2}), 128.5 (C^{A5}), 126.3 (C^{C4}), 124.2 (C^{A3}), 122.5 (C^{B3}). ESI-MS (MeOH, positive mode): m/z 990.90 [M − CF$_3$SO$_3$]$^+$ (calcd. 990.95), 421.60 [M − 2CF$_3$SO$_3$]$^{2+}$ (calcd. 421.00). UV-VIS (CH$_3$CN, 2.25 × 10^{-4} mol·dm^{-3}) λ_{max} (ε) 235 (25700). 256 (sh, 20900), 282 (sh, 32300), 289 (35300), 311 (sh, 25600), 329 (20700), 341 nm (16,800 dm^3·mol^{-1}·cm^{-1}). IR (see Figure S1). C$_{44}$H$_{28}$Br$_2$F$_6$N$_6$O$_6$S$_2$Zn requires C 46.36, H 2.48, N 7.37; found C 45.57, H 3.17, N 7.30%.

2.3. [Zn(**2**)$_2$][CF$_3$SO$_3$]$_2$

The method was as for [Zn(**1**)$_2$][CF$_3$SO$_3$]$_2$ but starting with Zn(CF$_3$SO$_3$)$_2$ (59.6 mg, 0.164 mmol) and **2** (445 mg, 0.337 mmol). [Zn(**2**)$_2$][CF$_3$SO$_3$]$_2$ was isolated as a white microcrystalline solid (190 mg, 0.152 mmol, 92.7%). X-ray quality crystals were grown by slow diffusion of Et$_2$O into an MeCN solution of [Zn(**2**)$_2$][CF$_3$SO$_3$]$_2$. ^{1}H NMR (CD$_3$CN, 500 MHz) δ/ppm: 9.00 (s, 4H, H^{B3}), 8.72 (dt, J = 8.1, 1.0 Hz, 4H, H^{A3}), 8.29 (m, 4H, H^{C2}), 8.18 (td, J = 7.8, 1.6 Hz, 4H, H^{A4}), 8.12 (m, 4H, H^{C3}), 7.84 (ddd, J = 5.1, 1.7, 0.9 Hz, 4H, H^{A6}), 7.41 (m, 4H, H^{A5}), 4.19 (m, 8H, H^{Et}), 1.36 (t, J = 7.0 Hz, 12H, H^{Et}). ^{13}C NMR (CD$_3$CN, 126 MHz) δ/ppm: 156.3 (C^{B4}), 151.0 (C^{B2}), 149.1 (C^{A6}), 148.7 (C^{A2}), 142.3 (C^{A4}), 140.9 (C^{C1}), 133.5 (d, J_{PC} = 9.9 Hz, C^{C3}), 132.0 (C^{C4}), 129.3 (d, J_{PC} = 14.9 Hz, C^{C2}), 128.6 (C^{A5}), 124 3 (C^{A3}), 123.0 (C^{B3}), 63.3 (d, J_{PC} = 5.6 Hz, C^{Et}), 16.7 (d, J_{PC} = 6.2 Hz, C^{Et}). ^{31}P{^{1}H} NMR (CD$_3$CN, 202 MHz) δ/ppm +16.3 (s). ESI-MS (MeOH, positive mode): m/z 1103.20 [M − CF$_3$SO$_3$]$^+$ (calcd. 1103.19), 477.65 [M − 2CF$_3$SO$_3$]$^{2+}$ (calcd. 477.12). UV-VIS (CH$_3$CN, 1.55 × 10^{-4} mol·dm^{-3}) λ_{max} (ε) 234 (23900), 266 (sh, 29100), 281 (sh, 43000), 288 (47500), 324 (15800), 342 nm (16,000 dm^3·mol^{-1}·cm^{-1}). IR (see Figure S2). C$_{52}$H$_{48}$F$_6$N$_6$O$_{12}$P$_2$S$_2$Zn requires C 49.79, H 3.86, N 6.70; found C 48.96, H 4.05, N 6.80%.

2.4. Crystallography

Single-crystal data were collected on a Bruker APEX-II diffractometer (Bruker AXS GmbH, Karlsruhe, Germany); data reduction, solution, and refinement used APEX2, SuperFlip, and CRYSTALS, respectively [35–37]. Structure analysis used Mercury v. 3.6 [38,39].

[Zn(**1**)$_2$][CF$_3$SO$_3$]$_2$·MeCN·1/2Et$_2$O: C$_{48}$H$_{36}$Br$_2$F$_6$N$_7$O$_{6.5}$S$_2$Zn, M = 1218.17, colourless block, triclinic, space group $P\bar{1}$, a = 12.1814(12), b = 14.7510(15), c = 15.1416(15) Å, α = 79.980(3), β = 69.658(3), γ = 71.317(3)°, U = 2410.6(4) Å^3, Z = 2, D_c = 1.678 Mg m^{-3}, μ(Cu-Kα) = 4.177 mm^{-1}, T = 123 K. Total 26609 reflections, 8560 unique, R_{int} = 0.029. Refinement of 8388 reflections (647 parameters) with $I > 2\delta(I)$ converged at final $R1$ = 0.0411 ($R1$ all data = 0.0416), $wR2$ = 0.0939 ($wR2$ all data = 0.0940), gof = 0.9775. CCDC 1879484.

[Zn(**2**)$_2$][CF$_3$SO$_3$]$_2$: C$_{52}$H$_{48}$F$_6$N$_6$O$_{12}$P$_2$S$_2$Zn, M = 1254.43, colourless block, monoclinic, space group $P2_1/n$, a = 15.7260(10), b = 15.6566(10), c = 22.9286(15) Å, β = 106.899(2)°, U = 5401.6(6) Å^3, Z = 4, D_c = 1.542 Mg m^{-3}, μ(Cu-Kα) = 2.693 mm^{-1}, T = 123 K. Total 69834 reflections, 9917 unique, R_{int} = 0.033. Refinement of 9379 reflections (730 parameters) with $I > 2\delta(I)$ converged at final $R1$ = 0.0372 ($R1$ all data = 0.0388), $wR2$ = 0.0923 ($wR2$ all data = 0.0924), gof = 1.0280. CCDC 1879482.

3. Results and Discussion

3.1. Synthesis and Solution Characterization of Complexes

Ligands **1** and **2** were prepared as previously reported [33,34]. Reaction of zinc(II) triflate with two equivalents of **1** or **2** resulted in the formation of $[Zn(1)_2][CF_3SO_3]_2$ and $[Zn(2)_2][CF_3SO_3]_2$ in 89.9% and 92.7% yield after purification. The electrospray mass spectra of the compounds showed peak envelopes with m/z 990.90 and 1103.20, respectively, arising from the $[M - CF_3SO_3]^+$ ion; the observed isotope distribution matched the simulation (Figures S3 and S4). The base peak in each spectrum corresponded to the $[M - 2CF_3SO_3]^{2+}$ ion with half-mass separations between the peaks in the peak-envelope. The ^1H and ^{13}C NMR spectra were assigned by 2D methods. Figures 1 and 2 display the aromatic regions of the ^1H NMR spectra, with the full spectra shown in Figures S5 and S6. COSY and HMQC spectra are shown in Figures S7–S10. The shift to lower frequency for the proton H^{A6} (see Scheme 2 for labelling) ongoing from free to coordinated ligand was typical of the formation of the {M(tpy)$_2$} unit, with H^{A6} lying over the π-system of the adjacent ligand. For example, in **2** (in CDCl$_3$), the resonance for H^{A6} appeared at δ 8.74 ppm [34], while in $[Zn(2)_2][CF_3SO_3]_2$ (in CD$_3$CN), it was at δ 7.88 ppm. In $[Zn(1)_2][CF_3SO_3]_2$, the ^1H NMR signal for H^{C2} showed a NOESY cross-peak to the signal for H^{B3} (Figure S11). Assignment in $[Zn(2)_2][CF_3SO_3]_2$ was also aided by characteristic coupling between the ^{31}P nucleus and protons H^{C2} and H^{C3}, respectively (Figure 2). $[Zn(2)_2][CF_3SO_3]_2$ exhibited a signal at δ 16.3 ppm in its ^{31}P NMR spectrum. The ethoxy groups in $[Zn(2)_2][CF_3SO_3]_2$ were characterized by multiplets at δ 4.19 and 1.36 ppm in the ^1H NMR spectrum and by doublets at δ 63.3 ppm (J_{PC} = 5.6 Hz) and δ 16.7 ppm (J_{PC} = 6.2 Hz). The IR spectra of $[Zn(1)_2][CF_3SO_3]_2$ and $[Zn(2)_2][CF_3SO_3]_2$ are shown in Figures S1 and S2. The most significant difference was an absorption at 959 cm^{-1} assigned to ν(P–O) in $[Zn(2)_2][CF_3SO_3]_2$.

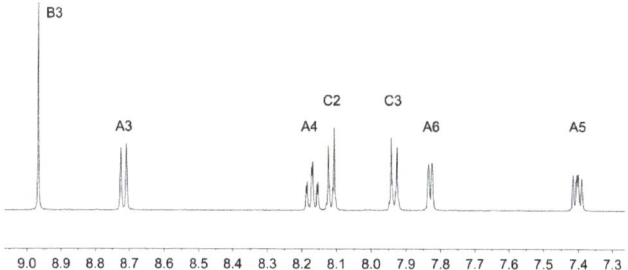

Figure 1. The ^1H NMR spectrum (500 MHz, CD$_3$CN) of $[Zn(1)_2][CF_3SO_3]_2$ (see also Figure S5). See Scheme 2 for atom labels.

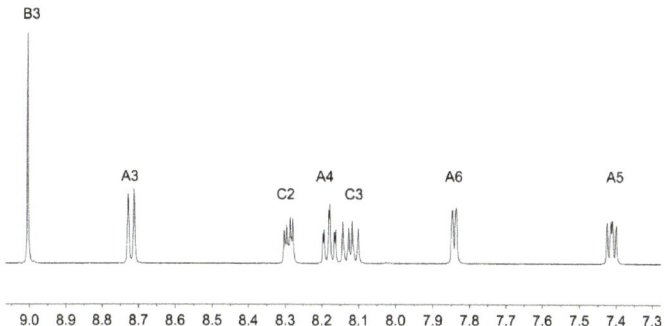

Figure 2. The ^1H NMR spectrum (500 MHz, CD$_3$CN) of $[Zn(2)_2][CF_3SO_3]_2$ (see also Figure S6). See Scheme 2 for atom labels.

3.2. Crystal Structures of [Zn(1)₂][CF₃SO₃]₂·MeCN·¹/₂Et₂O and [Zn(2)₂][CF₃SO₃]₂

Colourless single crystals of [Zn(1)₂][CF₃SO₃]₂·MeCN·¹/₂Et₂O and [Zn(2)₂][CF₃SO₃]₂ were grown by diffusion of Et₂O into MeCN solutions of the compounds. The two compounds crystallized in the triclinic P–1 and monoclinic P2₁/n space groups, respectively. Despite there being a wide range of crystallographically determined salts of [Zn(4′-Rtpy)₂]²⁺ (4′-Rtpy is a 4′-functionalized tpy), we were surprised to find no salts of [Zn(1)₂]²⁺ in the CSD (v. 5.39 with updates [24]. Figures 3 and 4 display the structures of the [Zn(1)₂]²⁺ and [Zn(2)₂]²⁺ cations, with selected bond distances and angles given in the captions to the figures. The Zn–N bond lengths were typical of {Zn(tpy)₂}²⁺ units, as were the N–Zn–N chelate angles. On going from the bromo to diethylphosphonate derivative, the tpy units become ruffled. In [Zn(1)₂]²⁺, the angles between the planes of adjacent pyridine rings in the two ligands were 12.7° and 3.5°, and 9.7° and 2.6°. In contrast, in [Zn(2)₂]²⁺, the corresponding angles were 8.0° and 19.0°, and 11.3° and 20.6°. Differences in the twist of the C₆H₄ ring with respect to the central pyridine of the tpy unit (21.0° and 32.3° for the two ligands in [Zn(1)₂]²⁺, and 30.6° and 34.0° in [Zn(2)₂]²⁺) were rationalized in terms of the different molecular packings discussed below.

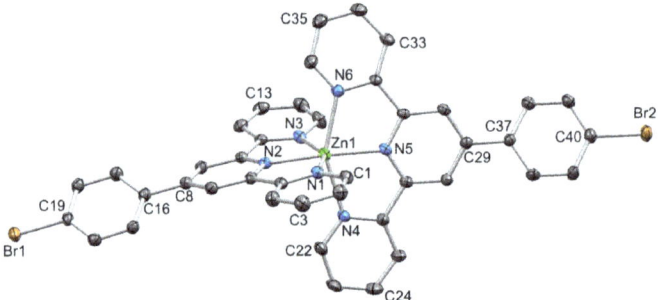

Figure 3. ORTEP-style depiction of the [Zn(1)₂]²⁺ cation. Hydrogen atoms omitted and ellipsoids plotted at 50% probability level. Selected bond parameters: Zn1–N1 = 2.179(2), Zn1–N2 = 2.087(2), Zn1–N3 = 2.198(2), Zn1–N4 = 2.212(2), Zn1–N5 = 2.079(2), Zn1–N6 = 2.169(2), C19–Br1 = 1.896(3), C40–Br2 = 1.900(3) Å; N2–Zn1–N5 = 174.89(9)°, N6–Zn1–N4 = 150.79(9)°, N2–Zn1–N1 = 75.46(9)°, N2–Zn1–N3 = 74.87(9)°, N5–Zn1–N4 = 75.22(9)°, N6–Zn1–N5 = 75.65(9)°.

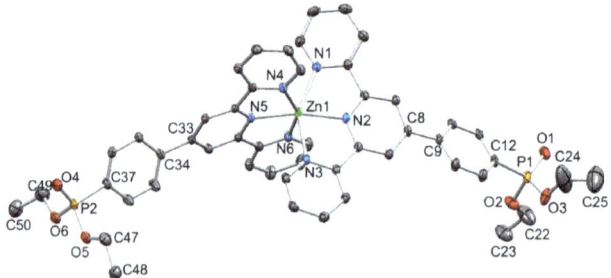

Figure 4. ORTEP-style diagram of the [Zn(2)₂]²⁺ cation. Ellipsoids plotted at 50% probability level; H atoms omitted. Selected bond parameters: Zn1–N1 = 2.2139(16), Zn1–N2 = 2.0760(14), Zn1–N3 = 2.2310(15), Zn1–N4 = 2.1701(15), Zn1–N5 = 2.0793(14), Zn1–N6 = 2.2157(15), P1–O1 = 1.4686(15), P1–O2 = 1.5713(15), P1–O3 = 1.5788(16), P2–O4 = 1.4683(15), P2–O5 = 1.5747(14), P2–O6 = 1.5829(16) Å; N2–Zn1–N5 = 167.72(6)°, N6–Zn1–N4 = 150.52(6)°, N2–Zn1–N1 = 75.28(5)°, N2–Zn1–N3 = 75.06(5)°, N5–Zn1–N4 = 76.23(6)°, N–Zn1–N6 = 74.75(6)°, O1–P1–O2 = 115.15(9)°, O1–P1–O3 = 115.54(9)°, O2–P1–O3 = 101.88(9)°, C12–P1–O1 = 114.01(9)°, C12–P1–O2 = 102.13(8)°, C12–P1–O3 = 106.55(9)°.

The packing of [Zn(1)$_2$]$^{2+}$ cations in [Zn(1)$_2$][CF$_3$SO$_3$]$_2$·MeCN·$^1/_2$Et$_2$O involved one of the characteristic 4′-Phtpy...4′-Phtpy (4′-Phtpy = 4′-phenyl-2,2′:6′,2″-terpyridine) embraces described by McMurtrie and Dance [40,41]. This is shown in Figure 5, with the embrace involving ligands containing Br2 and N6^i (and N6 and Br2^i, symmetry code $i = -x, 3 - y, 1 - z$). The phenyl$_{centroid}$...pyridine$_{centroid}$ separation was 3.73 Å and the angle between the π-stacked ring planes was 8.5°. For the ligand with Br2, the twist of the C$_6$H$_4$ ring with respect to the central pyridine of the tpy unit (21.0°) was smaller than for the ligand with Br1 (32.3°); this difference is associated with the π-stacking shown in Figure 5. By symmetry (in the P–1 space group), the ligand containing Br1 is also involved in a head-to-tail packing interaction with an adjacent ligand (Figure 5), but in this case, the relative positions of the symmetry-related ligands did not allow for an effective embrace [40,41]. Interestingly, there were no short Br...Br contacts, as have been observed for [Fe(1)$_2$][PF$_6$]$_2$ and related complexes [42]. We note that atom Br1 lies 4.13 Å from the centroid of the pyridine ring containing N4ii (symmetry code $ii = 2 - x, 2 - y, -z$) but at the upper limit of a weak Br...π contact [43]. The triflate ions are involved in F...H–C and O...H–C interactions. However, without crystallographic data for other salts, it was not possible to assess how the counterion affects the packing of the [Zn(1)$_2$]$^{2+}$ cations.

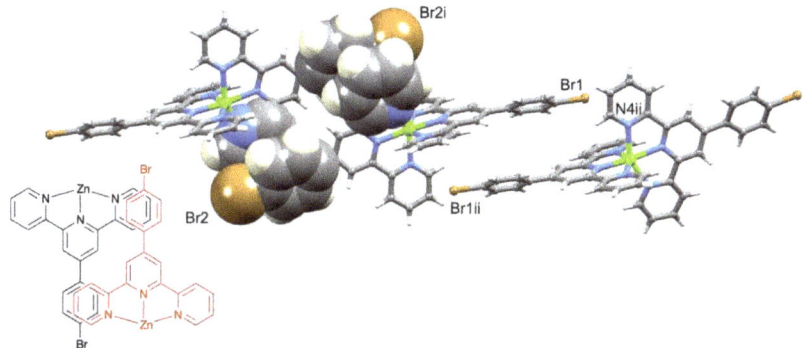

Figure 5. Packing interactions between centrosymmetric pairs of [Zn(1)$_2$]$^{2+}$ cations, and schematic representation of the 4′-Phtpy...4′-Phtpy embrace [37,38]. See text for symmetry codes.

A comparison of Figures 3 and 4 shows that going from bromo to phosphonate ester functionalities results in a significant "bowing" of the backbone of the {Zn(4′-XC$_6$H$_4$tpy)$_2$}$^{2+}$ unit, as is emphasized in the overlay in Figure 6. We have previously noted this feature in a number of compounds containing central {M(4′-XC$_6$H$_4$tpy)}$^{2+}$ motifs. However, detailed analysis of the packing interactions did not reveal a universal explanation for the phenomenon [44–46]. The bowing of the backbone in [Zn(2)$_2$]$^{2+}$ could be quantified by the C19...Zn1...C40 angle of 168.5° in [Zn(1)$_2$]$^{2+}$ (Figure 3) vs. the C12...Zn1...C37 (Figure 4) angle of 145.2° in [Zn(2)$_2$]$^{2+}$. [Zn(2)$_2$][CF$_3$SO$_3$]$_2$ crystallized without solvent in the lattice and solvent effects can be excluded as the cause of the deviation from an ideal C12...C37 vector in [Zn(2)$_2$]$^{2+}$. Important cation...cation contacts are illustrated in Figure 7. Pyridine rings containing N4 and N4^i engage in the face-to-face π-stacking interaction shown in Figure 7a with centroid...centroid and interplane separations of 3.63 and 3.24 Å, respectively. Symmetry-related intercation π-stacking contacts result in the formation of double-chains of cations running parallel to the crystallographic b-axis (Figure 7a). The bending of the backbone of the ligand incorporating atoms P1 and O1 appears to be associated with steric hindrance between the C$_6$H$_4$PO(OEt)$_2$ group and an adjacent pair of π-stacked pyridine rings. This also leads to close P–O...H–C contacts as shown in Figure 7b, parameters for which are O1...H271^{ii} = 2.43 Å, O1...H211^{iii} = 2.58 Å, and O1...H201^{iii} = 2.69 Å (symmetry codes are defined in the caption to Figure 7). Chains are further associated through short P–O...H–C contacts between phosphate ester groups of adjacent cations (Figure 8). This involves atom O5 (attached to P2) and H222^{iv} of the OCH$_2$ group attached to P1^{iv} (Figure 8b). Comparisons of the packing interactions in

[Zn(2)₂][CF₃SO₃]₂ with related structures are limited by the lack of structural information (see Section 1: Introduction). However, analysis of the structure of [ReCl{4'(4-(MeO)₂OPC₆H₄tpy}₂] (which contains a 7-coordinate Re(III) centre) revealed that the dominant features in the packing involved close OMe...π_{arene} contacts rather than tpy...tpy π-stacking (CSD refcode PACXOI) [32].

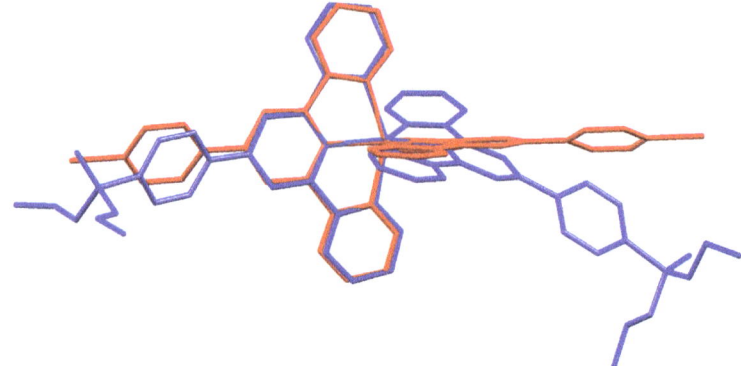

Figure 6. Overlay of the structures of the [Zn(1)₂]²⁺ (red) and [Zn(2)₂]²⁺ (blue) cations (H atoms omitted). Pairs of atoms Zn1, N2, N5, and N1 were superimposed to within ±0.13 Å.

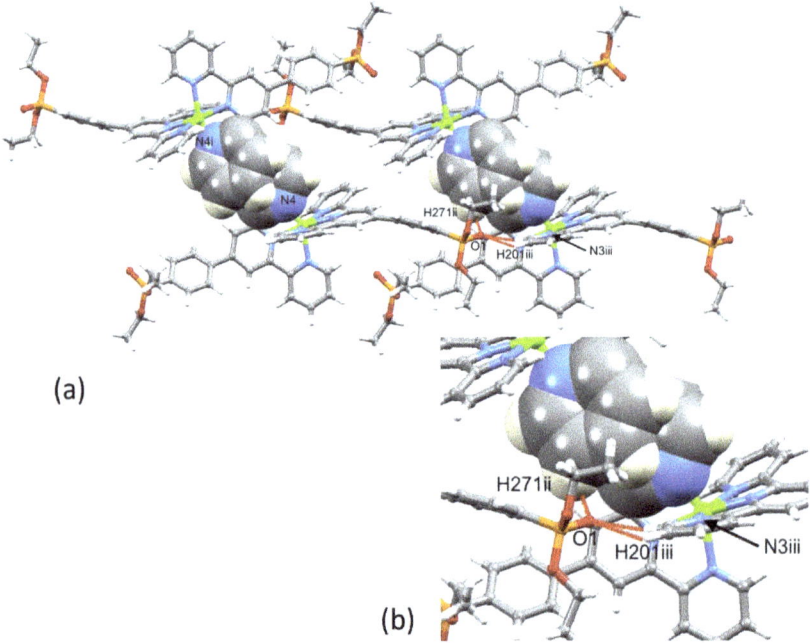

Figure 7. (a) Intercation packing interactions between [Zn(2)₂]²⁺ cations in [Zn(2)₂][CF₃SO₃]₂ and (b) expansion of the close P–O...H–C contacts. Anions are omitted. Symmetry codes: $i = -x, 1 - y, 1 - z$; $ii = -x, -y, 1 - z$; $iii = x, -1+y, +z$.

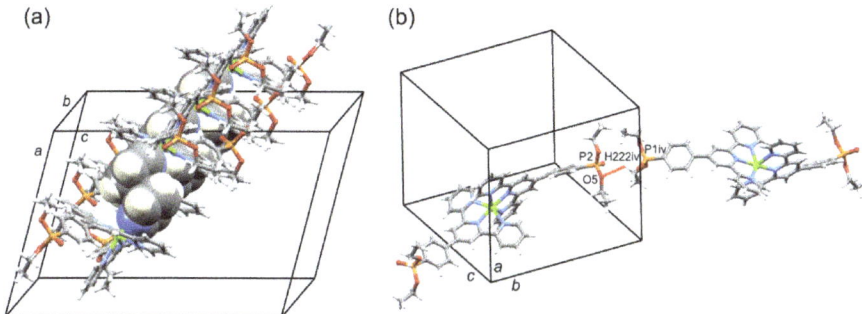

Figure 8. (a) Part of one chain of [Zn(2)$_2$]$^{2+}$ cations in [Zn(2)$_2$][CF$_3$SO$_3$]$_2$ supported by pyridine...pyridine π-stacking interactions. (b) Short PO...HC contact between phosphate ester groups of adjacent cations. Symmetry code iv = $^1/_2-x, ^3/_2+y, ^1/_2-z$.

4. Conclusions

We have prepared the bromo- and phosphonate-ester-functionalized bis(tpy) complexes [Zn(1)$_2$][CF$_3$SO$_3$]$_2$ and [Zn(2)$_2$][CF$_3$SO$_3$]$_2$, and electrospray mass spectrometric and multinuclear NMR spectroscopic data were consistent with the formation of the homoleptic complexes. The single-crystal structures of [Zn(1)$_2$][CF$_3$SO$_3$]$_2$·MeCN·$^1/_2$Et$_2$O and [Zn(2)$_2$][CF$_3$SO$_3$]$_2$ were determined and confirmed the expected octahedral coordination of the Zn^{2+} ion in each complex. Significant bowing of the backbone of the {Zn(4'-XC$_6$H$_4$tpy)$_2$}$^{2+}$ unit was observed ongoing from bromo to phosphonate ester functionalities and was associated with changes in packing interactions. The [Zn(1)$_2$]$^{2+}$ cations engaged in head-to-tail 4'-Phtpy...4'-Phtpy embraces with efficient pyridine...phenylene π-stacking interactions. In contrast, [Zn(2)$_2$]$^{2+}$ cations packed with one of the two ligands involved in pyridine...pyridine π-stacking; steric hindrance between one C$_6$H$_4$PO(OEt)$_2$ group and an adjacent pair of π-stacked pyridine rings resulted in distortion of the backbone of the ligand. Short P–O...H–C contacts between phosphate ester groups were also observed. This investigation was the first crystallographic determination of a salt of a homoleptic [M{4'-(RO)$_2$OPC$_6$H$_4$tpy}$_2$]$^{n+}$ cation.

Supplementary Materials: The following are available online at http://www.mdpi.com/2073-4352/8/12/461/s1. Figures S1 and S2: IR spectra of [Zn(1)$_2$][CF$_3$SO$_3$]$_2$ and [Zn(2)$_2$][CF$_3$SO$_3$]$_2$. Figures S3–S11: Additional NMR spectra of [Zn(1)$_2$][CF$_3$SO$_3$]$_2$ and [Zn(2)$_2$][CF$_3$SO$_3$]$_2$.

Author Contributions: D.Z.: Ligand and complex synthesis and characterization, consolidation of data; A.P.: Crystallography; E.C.C.: Group leader, project concepts, and contributions to manuscript preparation; C.E.H.: Group leader, project concepts, and manuscript preparation.

Funding: We thank the Swiss National Science Foundation and the University of Basel for support.

Conflicts of Interest: The authors declare no conflict of interest.

References

1. Constable, E.C. Homoleptic complexes of 2,2'-bipyridine. *Adv. Inorg. Chem.* **1989**, *34*, 1–63. [CrossRef]
2. Constable, E.C. The coordination chemistry of 2,2':6',2''-terpyridine and higher oligopyridines. *Adv. Inorg. Chem. Radiochem.* **1986**, *30*, 69–121. [CrossRef]
3. Constable, E.C. 2,2':6',2''-Terpyridines: From chemical obscurity to common supramolecular motifs. *Chem. Soc. Rev.* **2007**, *36*, 246–253. [CrossRef] [PubMed]
4. Schubert, U.S.; Hofmeier, H.; Newkome, G.R. *Modern Terpyridine Chemistry*; Wiley-VCH: Weinheim, Germany, 2006; ISBN 978-3-527-31475-1.
5. Zamalyutin, V.V.; Bezdenezhnykh, V.A.; Nichugovskiy, A.I.; Flid, V.R. New approaches to the synthesis of 2,2':6',2''-terpyridine and some of its derivatives. *Russ. J. Org. Chem.* **2018**, *54*, 419–425. [CrossRef]

6. Schubert, U.S.; Winter, A.; Newkome, G.R. *Terpyridine-Based Materials: For Catalyic, Optoelectronic and Life Sciene Applications*; Wiley-VCH: Weinheim, Germany, 2012; ISBN 978-3-527-33038-6.
7. Laschuk, N.O.; Ebralidze, I.I.; Poisson, J.; Egan, J.G.; Quaranta, S.; Allan, J.T.S.; Cusden, H.; Gaspari, F.; Naumkin, F.Y.; Easton, E.B.; et al. Ligand Impact on Monolayer Electrochromic Material Properties. *ACS Appl. Mater. Interfaces* **2018**, *10*, 35334–35343. [CrossRef] [PubMed]
8. Nazeeruddin, M.K.; Baranoff, E.; Grätzel, M. Dye-sensititzed solar cells: A brief review. *Sol. Energy* **2011**, *85*, 1172–1178. [CrossRef]
9. Aghazada, S.; Nazeeruddin, M.K. Ruthenium complexes as sensitisers in dye-sensititzed solar cells. *Inorganics* **2018**, *6*, 52. [CrossRef]
10. Zhang, L.; Cole, J.M. Anchoring groups for dye-sensitized solar cells. *ACS Appl. Mater. Interfaces* **2015**, *7*, 3427–3455. [CrossRef]
11. Marinakis, N.; Willgert, M.; Constable, E.C.; Housecroft, C.E. Optimization of performance and long-term stability of p-type dye-sensitized solar cells with a cycloruthenated dye through electrolyte solvent tuning. *Sustain. Energy Fuels* **2017**, *1*, 626–635. [CrossRef]
12. Marinakis, N.; Wobill, C.; Constable, E.C.; Housecroft, C.E. Refining the anchor: Optimizing the performance of cyclometallated ruthenium(II) dyes in p-type dye sensitized solar cells. *Polyhedron* **2018**, *140*, 122–128. [CrossRef]
13. Housecroft, C.E.; Constable, E.C. The emergence of copper(I)-based dye sensitized solar cells. *Chem. Soc. Rev.* **2015**, *44*, 8386–8398. [CrossRef] [PubMed]
14. Malzner, F.J.; Housecroft, C.E.; Constable, E.C. The versatile SALSAC approach to heteroleptic copper(I) dye assembly in dye-sensitized solar cells. *Inorganics* **2018**, *6*, 57. [CrossRef]
15. Bozic-Weber, B.; Constable, E.C.; Hostettler, N.; Housecroft, C.E.; Schmitt, R.; Schönhofer, E. The d^{10} route to dye-sensitized solar cells: Step-wise assembly of zinc(II) photosensitizers on TiO_2 surfaces. *Chem. Commun.* **2012**, *48*, 5727–5729. [CrossRef] [PubMed]
16. Hostettler, N.; Fürer, S.O.; Bozic-Weber, B.; Constable, E.C.; Housecroft, C.E. Alkyl chain-functionalized hole-transporting domains in zinc(II) dye-sensitized solar cells. *Dyes Pigments* **2015**, *116*, 124–130. [CrossRef]
17. Bozic-Weber, B.; Brauchli, S.Y.; Constable, E.C.; Fürer, S.O.; Housecroft, C.E.; Malzner, F.J.; Wright, I.A.; Zampese, J.A. Improving the photoresponse of copper(I) dyes in dye-sensitized solar cells by tuning ancillary and anchoring ligand modules. *Dalton Trans.* **2013**, *42*, 12293–12308. [CrossRef] [PubMed]
18. Stergiopoulos, T.; Arabatzis, I.M.; Kalbac, M.; Lukes, I.; Falaras, P. Incorporation of innovative compounds in nanostructured photoelectrochemical cells. *J. Mater. Process. Technol.* **2005**, *161*, 107–112. [CrossRef]
19. Stergiopoulos, T.; Bernard, M.-C.; Hugot-Le Goff, A.; Falaras, P. Resonance micro-Raman spectrophotoelectrochemistry on nanocrystalline TiO_2 thin film electrodes sensitized by Ru (II) complexes. *Coord. Chem. Rev.* **2004**, *248*, 1407–1420. [CrossRef]
20. Krebs, F.C.; Biancardo, M. Dye sensitized photovoltaic cells: Attaching conjugated polymers to zwitterionic ruthenium dyes. *Solar Energy Mater. Solar Cells* **2006**, *90*, 142–165. [CrossRef]
21. Jing, B.; Zhang, H.; Zhang, M.; Lu, Z.; Shen, T. Ruthenium(II) thiocyanate complexes containing 4'-(4-phosphonatophenyl)-2,2':6',2''-terpyridine: Synthesis, photophysics and photosensitization to nanocrystalline TiO_2 electrodes. *J. Mater. Chem.* **1998**, *8*, 2055–2060. [CrossRef]
22. Wolpher, H.; Sinha, S.; Pan, J.; Johansson, A.; Lundqvist, M.J.; Persson, P.; Lomoth, R.; Bergquist, J.; Sun, L.; Sundström, V.; et al. Synthesis and electron transfer studies of ruthenium–terpyridine-based dyads attached to nanostructured TiO_2. *Inorg. Chem.* **2007**, *46*, 638–651. [CrossRef]
23. Lundqvist, M.J.; Nilsig, M.; Lunell, S.; Åkermark, B.; Persson, P. Spacer and anchor effects on the electronic coupling in ruthenium-bis-terpyridine dye-sensitized TiO_2 nanocrystals studied by DFT. *J. Phys. Chem. B* **2006**, *110*, 20513–20525. [CrossRef] [PubMed]
24. Groom, C.R.; Bruno, I.J.; Lightfoot, M.P.; Ward, S.C. The Cambridge Structural Database. *Acta Crystallogr. Sect. B* **2016**, *72*, 171–179. [CrossRef] [PubMed]
25. Laschuk, N.O.; Ebralidze, I.I.; Spasyuk, D.; Zenkina, O.V. Multi-readout logic gate for the selective detection of metal ions at the parts per billion level. *Eur. J. Inorg. Chem.* **2016**, 3530–3535. [CrossRef]

26. Constable, E.C.; Housecroft, C.E.; Neuburger, M.; Schneider, A.G.; Zehnder, M. The ditopic ligands 4′-(diphenylphosphino)-2,2′:6′,2″-terpyridine (**1**) and 4′-(oxodiphenylphosphanyl)-2,2′:6′,2″-terpyridine (**2**): Coordination to iron(II), ruthenium(II), cobalt(II) and palladium(II); X-ray crystal structures of [Ru(**2**)$_2$][PF$_6$]$_2$·MeCN·H$_2$O and *trans*-[Pd(**1**)$_2$Cl$_2$]·2.5CH$_2$Cl$_2$. *J. Chem. Soc. Dalton Trans.* **1997**, 2427–2434. [CrossRef]
27. Chen, W.; Rein, F.N.; Scott, B.L.; Rocha, R.C. (2,2′-Bipyridine)chlorido[diethyl(2,2′:6′,2″-terpyridin-4-yl) phosphonate]ruthenium(II) hexafluoridophosphate acetonitrile/water solvate. *Acta Crystallogr. Sect. E* **2013**, *69*, m510–m511. [CrossRef] [PubMed]
28. Zakeeruddin, S.M.; Nazeeruddin, M.K.; Pechy, P.; Rotzinger, F.P.; Humphry-Baker, R.; Kalyanasundaram, K.; Grätzel, M.; Shklover, V.; Haibach, T. Molecular engineering of photosensitizers for nanocrystalline solar cells: Synthesis and characterization of Ru dyes based on phosphonated terpyridines. *Inorg. Chem.* **1997**, *36*, 5937–5946. [CrossRef] [PubMed]
29. Francàs, L.; Richmond, C.; Garrido-Barros, P.; Planas, N.; Roeser, S.; Benet-Buchholz, J.; Escriche, L.; Sala, X.; Llobet, A. Ru–bis(pyridine)pyrazolate (bpp)-based water-oxidation catalysts anchored on TiO$_2$: The importance of the nature and position of the anchoring group. *Chem. Eur. J.* **2016**, *22*, 5261–5268. [CrossRef]
30. Ferrer, I.; Fontrodona, X.; Roig, A.; Rodríguez, M.; Romero, I. A recoverable ruthenium aqua complex supported on silica particles: An efficient epoxidation catalyst. *Chem. Eur. J.* **2017**, *23*, 4096–4107. [CrossRef] [PubMed]
31. Vaquer, L.; Riente, P.; Sala, X.; Jansat, S.; Benet-Buchholz, J.; Llobet, A.; Pericas, M.A. Molecular ruthenium complexes anchored on magnetic nanoparticles that act as powerful and magnetically recyclable stereospecific epoxidation catalysts. *Cat. Sci. Tech.* **2013**, *3*, 706–714. [CrossRef]
32. Sugimoto, H.; Tano, H.; Miyake, H.; Itoh, S. Seven-coordinate rhenium(III) complexes with a labile coordination site assembled on indium-doped tin oxide (ITO) electrodes: Catalytic reduction of dioxygen to hydrogen peroxide. *Chem. Lett.* **2010**, *39*, 986–988. [CrossRef]
33. Wang, J.; Hanan, G.S. A facile route to sterically hindered and non-hindered 4′-aryl-2,2′:6′,2″-terpyridines. *Synlett* **2005**, 1251–1254. [CrossRef]
34. Spampinato, V.; Tuccitto, N.; Quici, S.; Calabrese, V.; Marletta, G.; Torrisi, A.; Licciardello, A. Functionalization of oxide surfaces by terpyridine phosphonate ligands: Surface reactions and anchoring geometry. *Langmuir* **2010**, *26*, 8400–8406. [CrossRef] [PubMed]
35. Bruker Analytical X-ray Systems. *APEX2, Version 2 User Manual*; M86-E01078; Bruker Analytical X-ray Systems, Inc.: Madison, WI, USA, 2006.
36. Palatinus, L.; Chapuis, G. SUPERFLIP—A computer program for the solution of crystal structures by charge flipping in arbitrary dimensions. *J. Appl. Cryst.* **2007**, *40*, 786–790. [CrossRef]
37. Betteridge, P.W.; Carruthers, J.R.; Cooper, R.I.; Prout, K.; Watkin, D.J. CRYSTALS Version 12: Software for Guided Crystal Structure Analysis. *J. Appl. Cryst.* **2003**, *36*, 1487. [CrossRef]
38. Macrae, C.F.; Edgington, P.R.; McCabe, P.; Pidcock, E.; Shields, G.P.; Taylor, R.; Towler, M.; van de Streek, J. Mercury: Visualization and analysis of crystal structures. *J. Appl. Cryst.* **2006**, *39*, 453–457. [CrossRef]
39. Macrae, C.F.; Bruno, I.J.; Chisholm, J.A.; Edgington, P.R.; McCabe, P.; Pidcock, E.; Rodriguez-Monge, L.; Taylor, R.; van de Streek, J.; Wood, P.A. Mercury CSD 2.0—New Features for the Visualization and Investigation of Crystal Structures. *J. Appl. Cryst.* **2008**, *41*, 466–470. [CrossRef]
40. McMurtrie, J.; Dance, I. Crystal packing in metal complexes of 4′-phenylterpyridine and related ligands: Occurrence of the 2D and 1D terpy embrace arrays. *CrystEngComm* **2009**, *11*, 1141–1149. [CrossRef]
41. McMurtrie, J.; Dance, I. Alternative two-dimensional embrace nets formed by metal complexes of 4′-phenylterpyridine crystallised with hydrophilic anions. *CrystEngComm* **2010**, *12*, 3207–3217. [CrossRef]
42. Medleycott, E.A.; Hanan, G.S.; Abedin, T.S.M.; Thompson, L.K. The effect of steric hindrance on the Fe(II) complexes of triazine-containing ligands. *Polyhedron* **2008**, *27*, 493–501. [CrossRef]
43. Shukla, R.; Panini, P.; McAdam, C.J.; Robinson, B.H.; Simpson, J.; Tagg, T.; Chopra, D. Characterization of non-classical C–Br...π interactions in (E)-1,3-dibromo-5-(2-(ferrocenyl)vinyl)benzene and related derivatives of ferocene. *J. Mol. Struct.* **2017**, *131*, 16–24. [CrossRef]
44. Constable, E.C.; Housecroft, C.E.; Neuburger, M.; Schaffner, S.; Schaper, F. The solid-state structure of bis(4′-(4-pyridyl)-2,2′:6′,2″-terpyridine)ruthenium hexafluorophosphate nitrate—An expanded 4,4′-bipyridine. *Inorg. Chem. Commun.* **2006**, *9*, 616–619. [CrossRef]

45. Beves, J.E.; Constable, E.C.; Housecroft, C.E.; Kepert, C.J.; Neuburger, M.; Price, D.J.; Schaffner, S. The conjugate acid of bis{4′-(4-pyridyl)-2,2′:6′,2″-terpyridine}iron(II) as a self-complementary hydrogen-bonded building block. *CrystEngComm* **2007**, *9*, 1073–1077. [CrossRef]
46. Beves, J.E.; Bray, D.J.; Clegg, J.K.; Constable, E.C.; Housecroft, C.E.; Jolliffe, K.A.; Kepert, C.J.; Lindoy, L.F.; Neuburger, M.; Price, D.J.; et al. Expanding the 4,4′-bipyridine ligand: Structural variation in {M(pytpy)$_2$}$^{2+}$ complexes (pytpy = 4′-(4-pyridyl)-2,2′:6′,2″-terpyridine, M = Fe, Ni, Ru) and assembly of the hydrogen-bonded, one-dimensional polymer {[Ru(pytpy)(Hpytpy)]}$_n$$^{3n+}$. *Inorg. Chim. Acta* **2008**, *361*, 2582–2590. [CrossRef]

© 2018 by the authors. Licensee MDPI, Basel, Switzerland. This article is an open access article distributed under the terms and conditions of the Creative Commons Attribution (CC BY) license (http://creativecommons.org/licenses/by/4.0/).

Perspective

New Directions in Metal Phosphonate and Phosphinate Chemistry

Stephen J.I. Shearan [1], Norbert Stock [2], Franziska Emmerling [3], Jan Demel [4], Paul A. Wright [5], Konstantinos D. Demadis [6], Maria Vassaki [6], Ferdinando Costantino [7], Riccardo Vivani [8], Sébastien Sallard [9], Inés Ruiz Salcedo [10], Aurelio Cabeza [10] and Marco Taddei [1,*]

1. Energy Safety Research Institute, Swansea University, Fabian Way, Swansea SA1 8EN, UK; 940872@swansea.ac.uk
2. Institute of Inorganic Chemistry, Christian-Albrechts-University, Max-Eyth-Str. 2, 24118 Kiel, Germany; stock@ac.uni-kiel.de
3. Federal Institute for Materials Research and Testing (BAM), Richard-Willstaetter-Str. 11, 12489 Berlin, Germany; franziska.emmerling@bam.de
4. Institute of Inorganic Chemistry of the Czech Academy of Sciences, Husinec-Řež 1001, 250 68 Řež, Czech Republic; demel@iic.cas.cz
5. EaStCHEM School of Chemistry, University of St Andrews, Purdie Building, North Haugh, St Andrews KY16 9ST, UK; paw2@st-andrews.ac.uk
6. Crystal Engineering, Growth, and Design Laboratory, Department of Chemistry, University of Crete, Crete GR-71003 Heraklion, Greece; demadis@uoc.gr (K.D.D.); vassakimar@gmail.com (M.V.)
7. Department of Chemistry, Biology and Biotechnologies, Via Elce di Sotto n. 8, 06123 Perugia, Italy; ferdinando.costantino@unipg.it
8. Department of Pharmaceutical Sciences, University of Perugia, Via del Liceo 1, 06123 Perugia, Italy; riccardo.vivani@unipg.it
9. Flemish Institute for Technological Research—VITO, Sustainable Materials Department, Boeretang 200, 2400 Mol, Belgium; sebastien.sallard@vito.be
10. Depto. Química Inorgánica, Cristalografía y Mineralogía, Campus de Teatinos s/n, Universidad de Málaga, 29071 Málaga, Spain; inesrs@uma.es (I.R.S.); aurelio@uma.es (A.C.)
* Correspondence: marco.taddei@swansea.ac.uk; Tel.: +44-(0)1792-606230

Received: 30 April 2019; Accepted: 21 May 2019; Published: 24 May 2019

Abstract: In September 2018, the First European Workshop on Metal Phosphonates Chemistry brought together some prominent researchers in the field of metal phosphonates and phosphinates with the aim of discussing past and current research efforts and identifying future directions. The scope of this perspective article is to provide a critical overview of the topics discussed during the workshop, which are divided into two main areas: synthesis and characterisation, and applications. In terms of synthetic methods, there has been a push towards cleaner and more efficient approaches. This has led to the introduction of high-throughput synthesis and mechanochemical synthesis. The recent success of metal–organic frameworks has also promoted renewed interest in the synthesis of porous metal phosphonates and phosphinates. Regarding characterisation, the main advances are the development of electron diffraction as a tool for crystal structure determination and the deployment of in situ characterisation techniques, which have allowed for a better understanding of reaction pathways. In terms of applications, metal phosphonates have been found to be suitable materials for several purposes: they have been employed as heterogeneous catalysts for the synthesis of fine chemicals, as solid sorbents for gas separation, notably CO_2 capture, as materials for electrochemical devices, such as fuel cells and rechargeable batteries, and as matrices for drug delivery.

Keywords: metal phosphonates and phosphinates; layered materials; metal–organic frameworks; synthesis; X-ray and electron diffraction; in situ characterisation; heterogeneous catalysis; gas sorption/separation; proton conduction; rechargeable batteries; drug delivery

1. Introduction

Metal phosphonates (MPs) are a class of inorganic–organic hybrid polymeric materials built by the coordination of phosphonate ligands to metal ions, forming extended structures of various dimensionalities [1]. The field of MPs chemistry has seen steady growth over the last few decades, which has been driven by the interest for applications in areas such as ion exchange [2], intercalation chemistry [3–5], proton conduction [6], catalysis [7], and others. This is principally due to the exceptional chemical and thermal stability and high insolubility of these materials in many solvents, which can be attributed to the hard character of the phosphonate oxygen atoms and their high coordination affinity for metal atoms. MPs chemistry turned 40 years old in 2018. In September 2018, the First European Workshop on Metal Phosphonates Chemistry was held at the Energy Safety Research Institute of Swansea University and attended by several researchers in the field of MPs based in Europe. The idea behind the event was to bring these researchers together for the first time and create a forum for discussion about the state of the art in the field and promote future collaborations. This article is intended to highlight some of the most promising new avenues of research discussed during the First European Workshop on Metal Phosphonates Chemistry. As such, we are not seeking to provide an exhaustive review of the recent progresses in the field, which have been covered by several other publications over the last few years [6,8–14], including a comprehensive book published in 2011 [1].

Since the field of MPs has come a long way, this article starts off by looking into the past and retracing some of the most important stages of early development, with a focus on structural chemistry. The third section is devoted to synthesis and characterisation methods, placing emphasis on innovative approaches such as high-throughput synthesis for the discovery of new compounds, mechanochemical synthesis, strategies to form porous frameworks, structure solution from electron diffraction, and in situ characterisation methods. The fourth section focuses on novel applications of MPs, including gas sorption/separation, catalysis, electrochemical devices (fuel cells and rechargeable batteries), and drug delivery. Finally, we take a look at what the future of the field might hold, trying to identify the most promising new directions for innovative research.

2. Historical Landmarks

The initial interest in the field stemmed from the work by Clearfield et al. in the field of tetravalent metal phosphates, which have been known for their ion exchange properties since the 1950s. The key step forward made by Clearfield was the determination of the crystal structure of α-zirconium bis(monohydrogen orthophosphate) monohydrate [$Zr(HPO_4)_2 \cdot H_2O$, hereafter α-ZrP] [15] (Figure 1a,b) from single-crystal X-ray diffraction (SCXRD) data in 1968. α-ZrP has a layered structure constituted of Zr atoms octahedrally coordinated by tridentate monohydrogenphosphate groups, thus leaving free –OH groups pointing towards the interlayer space and hydrogen bonding to water molecules accommodated between the layers. The atomic level understanding of the structure of α-ZrP triggered intense research that aimed at taking advantage of the acidic protons on the surface of the layers, especially for ion exchange and intercalation purposes.

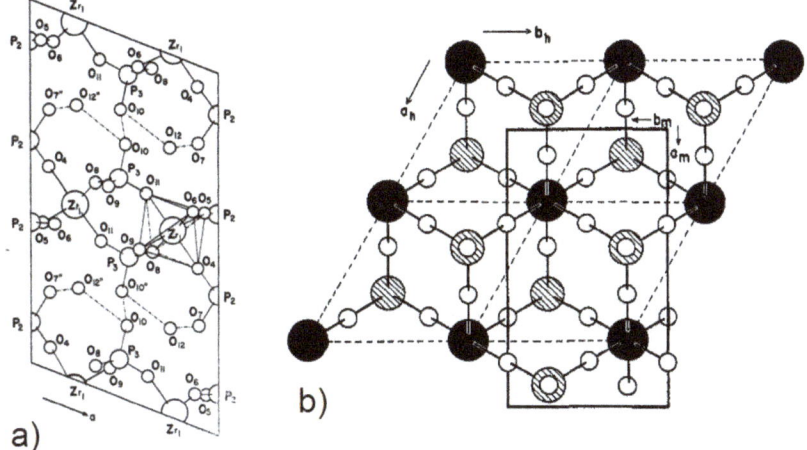

Figure 1. Crystal structure of α-ZrP viewed along the *b*-axis (**a**) and perpendicular to the layer (**b**). Adapted with permission from reference. [15]. Copyright 1969, American Chemical Society.

It was Alberti and Costantino et al. in 1978 [16] that would inaugurate the field of MPs. They prepared three zirconium phosphonates: zirconium phenylphosphonate, $Zr(C_6H_5PO_3)_2$; hydroxymethylphosphonate, $Zr(HOCH_2PO_3)_2$; and ethylphosphate, $Zr(C_2H_5OPO_3)_2$. Due to the very high insolubility of these compounds, they were unable to obtain suitable single crystals for solving the structures by SCXRD. Given the similar method of preparation to that of α-ZrP and based on the powder X-ray diffraction (PXRD) patterns dominated by intense basal peaks at low angles, the authors hypothesised that these compounds would give comparable layered structures to α-ZrP, where the –OH groups in the interlayer are substituted by the organic moieties, which determine the interlayer distance (Figure 2). In the following years, a number of similar compounds were prepared, based on both monophosphonates and diphosphonates, obtaining analogous layered or pillared-layered structures, respectively [17,18].

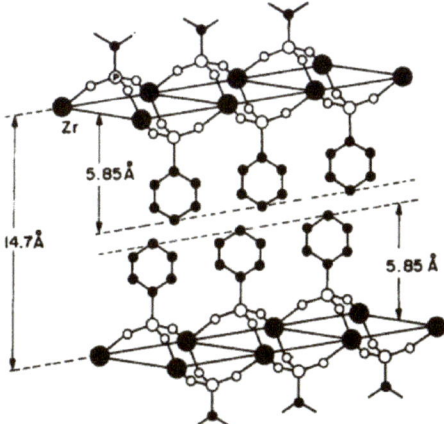

Figure 2. Idealised structural model of $Zr(C_6H_5PO_3)_2$. Reprinted with permission from reference [16]. Copyright 1978, Elsevier Inc.

The initial MP compounds were based on tetravalent metals, especially zirconium, but the late 1970s and 1980s brought about several structures based on divalent metal ions. Early work by Cunningham et al. [19] looked at divalent metal phenylphosphonates and phenylarsonates, M($C_6H_5PO_3$) and (M($C_6H_5AsO_3$), with M^{2+} = Mg, Mn Fe, Co, Ni, Cu, Zn, and Cd. They found that most of the synthetic processes were straightforward, and metal phenylphosphonates could be obtained by a simple reaction with the chloride or sulphate metal salts, with the exception of magnesium and iron. Thanks to the lower insolubility of these compounds, compared to tetravalent MPs, single crystals could be grown. Towards the end of the 1980s, a number of key papers detailing the crystal structures of various divalent MPs appeared [20–22]. All of these compounds featured structures based on layers built by the connection of metal atoms and phosphonate groups, with the organic moiety accommodated in the interlayer space. Figure 3 shows the layered crystal structure of Mn($C_6H_5PO_3$)·H_2O, which is just one of the examples provided by Cao et al. (1988) [21], where a number of structures based on divalent metals were described, based on SCXRD data, including: Mn, Mg, Ca, Cd, and Zn. Moving into the 1990s, MP structures expanded throughout the periodic table, covering a range of transition metals, all of the lanthanide series, and more than half of the s-block elements [1].

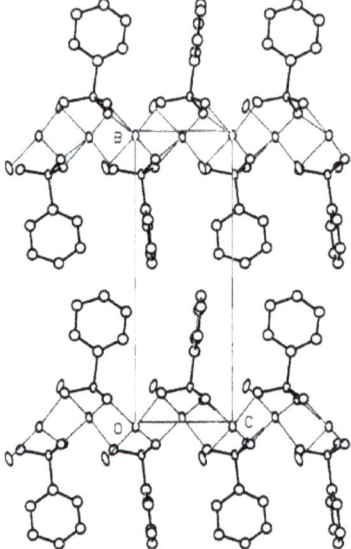

Figure 3. Structure of Mn($O_3PC_6H_5$)·H_2O viewed down the *a*-axis. Reprinted with permission from reference [21]. Copyright 1988, American Chemical Society.

While this expansion across the periodic table was taking place, there was also significant progress made with solving the crystal structures from PXRD data. This was of great importance for the field of MPs, since it has often proven difficult to obtain suitable single crystals for SCXRD, especially when metals with high oxidation states were involved. An example that has already been noted comes from the pioneering work of Alberti and Costantino et al. [16], who were unable to obtain large crystals for Zr($C_6H_5PO_3$)$_2$. The structure was eventually solved by Poojary et al. in 1993 [23] from PXRD data, using a combination of modelling techniques and Patterson methods, and refined by Rietveld methods, confirming that the assumption made by Alberti and Costantino et al. 15 years before was indeed correct and the structure of Zr phenylphosphonate was based on the same layered arrangement of α-ZrP (Figure 4a,b). With the development of more sophisticated laboratory powder X-ray diffractometers, the increased accessibility of synchrotron sources, and the availability of more powerful crystallographic software, structural solutions from PXRD have progressively become a workhorse for researchers working with MPs.

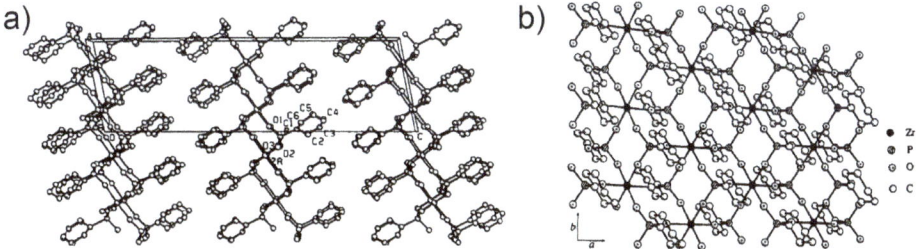

Figure 4. (**a**) Crystal structure of Zr(C$_6$H$_5$PO$_3$)$_2$ viewed along the *b*-axis, and (**b**) and perpendicular to the layer. Reprinted with permission from reference [23]. Copyright 1993, International Union of Crystallography.

While the overwhelming majority of MPs described in the first 15 years of research displayed layered structures, the early to mid-1990s witnessed the discovery of several open framework MPs [24–26]. The first examples were all based on the small ligand methylphosphonic acid, which afforded structures with channel-like arrangement, which was reminiscent of some zeolite frameworks, when combined with Cu, Zn, and Al (Figure 5a–c) [24–28]. The structural analogy with zeolites (and porous aluminium phosphates), which were at the time the most important class of crystalline and microporous materials, fuelled further investigation in the same direction [29]. This led to the discovery of more open framework materials, which were typically based on either monophosphonates or diphosphonates with short alkyl chains [30,31]. Increasing the length of the alkyl chains almost exclusively produced layered or pillared-layered structures, thus preventing the expansion of the channel size and generation of materials with higher porosity.

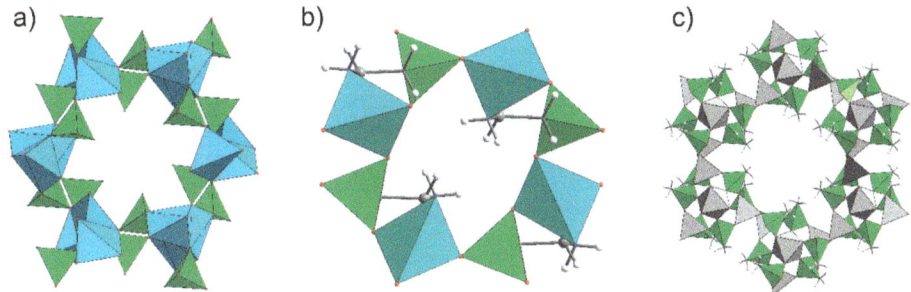

Figure 5. Crystal structures of the open framework metal phosphonates (MPs): (**a**) β-Cu(CH$_3$PO$_3$), (**b**) Zn(O$_3$PC$_2$H$_4$NH$_2$), and (**c**) α-Al$_2$(O$_3$PCH$_3$)$_2$. Colour code: light blue: copper, turquoise: zinc, light grey: aluminium, green: phosphorus, red: oxygen, blue: nitrogen, grey: carbon, white: hydrogen.

At the other end of the dimensionality scale, molecular MPs also started being the object of research in the second half of the 1990s, primarily for their potential as single molecule magnets and as model compounds for phosphate materials, especially those of group 13 elements (Figure 6) [9,32–36]. In order to overcome the strong tendency of MPs to polymerise, three strategies were devised to prevent the expansion of the structure: (1) using terminal ancillary ligands that can occupy coordination sites on the metal ions; (2) introducing sterically demanding groups on the phosphonic acids' backbone; (3) using a preformed cluster and performing controlled ligand exchange [9].

The reader interested in a more comprehensive account about the progress of the field until 2011 can refer to the book "Metal Phosphonate Chemistry: from Synthesis to Applications" [1], which also includes a chapter on the early history of MPs chemistry, authored by Abraham Clearfield [37].

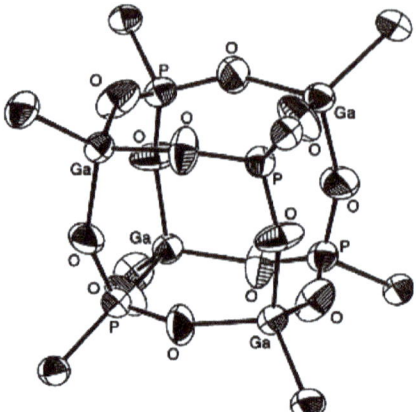

Figure 6. Crystal structure of the $Ga_4P_4O_{12}$ core found in $[^tBuGa(\mu_3-O_3PPh)]_4$. Reprinted with permission from reference [32]. Copyright 2003, Wiley-VCH Verlag GmbH & Co.

3. Synthesis and Characterisation of Metal Phosphonates

3.1. High-Throughput Methods

The synthesis of MPs is usually performed under hydrothermal or solvothermal conditions, with reaction times in the range of a few hours up to several days [38]. Identifying the proper conditions to obtain crystalline products involves the screening of reaction parameters such as pH, temperature, the concentration and molar ratios of reagents, and the amount of mineralisers or crystallisation modulators, which makes the discovery of novel MPs a time-consuming process. In addition, each "metal phosphonate" reaction system is unique, with its own idiosyncrasies. High-throughput (HT) methods have long been used to screen for ideal synthesis parameters, often leading to the accelerated discovery of novel compounds and further optimisation for increased yield [39] (Figure 7). This is also true for their application in MP synthesis, with the first reports of this approach falling in the early 2000s [40–44], and the methods are nowadays routinely used [45–47]. The interest in HT methods stems from their high efficiency, allowing for the systematic investigation of reaction parameters.

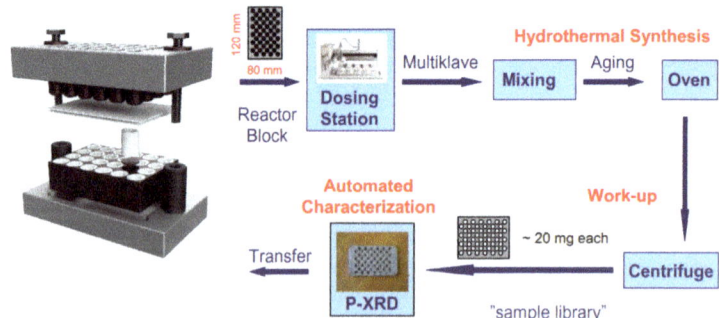

Figure 7. Typical workflow for a high-throughput (HT) synthesis experiment. Adapted with permission from reference [38]. Copyright 2009, Elsevier Inc.

Some of the earliest work on HT methods applied to MPs was carried out in 2004 [40], and looked at MPs based on the ligands $(H_2O_3PCH_2)_2N(CH_2)_4N(CH_2PO_3H_2)_2 \cdot 2H_2O$ ($H_8L^1 \cdot 2H_2O$), and p-$H_2O_3PCH_2C_6H_4COOH$ (H_3L^2). The investigation began with the design of two experiments to look at the system $Zn(NO_3)_2/H_3L^2/NaOH$. The first of the two varied the molar concentrations of the three starting materials while keeping the water content the same (Figure 8). This led to the discovery of three single crystalline compounds $Zn(HO_3PCH_2C_6H_4COOH)_2$ (1), $Zn(O_3PCH_2C_6H_4COOH)$ (2), and $Zn_3(HO_3PCH_2C_6H_4COO)_2 \cdot 4H_2O$ (3). Throughout the experiment, pH measurements were taken periodically, allowing the researchers to correlate the pH measurements of the reaction mixture with the dimensionality of the MP structure. Thus (3), featuring a three-dimensional (3D) framework structure based on highly condensed inorganic units, was favoured at pH ≈ 6, whereas (1) (pH ≈ 0) and (2) (pH ≈ 1) displayed isolated columns and layered structures, respectively. The second experiment used the opposite parameters to the first, with the molar ratio for $Zn(NO_3)_2/H_3L^2/NaOH$ kept constant while adjusting the water content. This led to the discovery of a fourth phase, $Zn(HO_3PCH_2C_6H_4COOH)_2 \cdot 4H_2O$ (4).

The researchers also carried out an investigation of the system $M^{2+}/H_8L^1 \cdot 2H_2O$ based on previous results [48]. In this case, various divalent ions were employed: M = Mg, Ca, Sr, Ba, Mn, Fe, Co, Ni, Zn, Cd, Sn, and Pb. As with the Zn-phosphonates, the investigation was split between two experiments, with the first taking the form of a discovery library for which the molar ratios for the starting materials are adjusted (e.g., 1:1, 1:4, 4:1, 2:3), while the water content was kept constant. Results showed that the microcrystalline products had formed for M = Mg, Ca, Mn, Fe, Co, Ni, Zn, and Cd, and that the compounds were isostructural. The second experiment showed that it was also possible to obtain single crystals, where M = Mg, Mn, Co, Ni, Cd, and Zn, by keeping the $M^{2+}:H_8L^1$ molar ratio at 1:1 and adjusting the water content between 98–99.83 mol % in order to tune the concentration of the starting materials. In total, 48 individual reactions were carried out without direct manipulation of each sample, while also exploring a large parameter space.

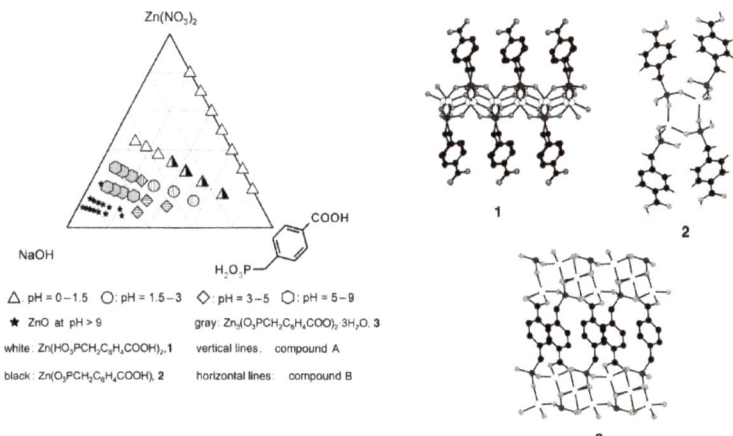

Figure 8. Crystallization diagram of observed phases in the $Zn(NO_3)_2/p$-$H_2O_3PCH_2C_6H_4COOH/NaOH$ system (left). Sections of the structures of $Zn(HO_3PCH_2C_6H_4COOH)_2$ (1), $Zn(O_3PCH_2C_6H_4COOH)$ (2), and $Zn_3(HO_3PCH_2C_6H_4COO)_2 \cdot 4H_2O$ (3) (right). Colour code: Zn: white, C: black, P: dark grey, O: light grey. Reprinted with permission from reference. [40]. Copyright 2004, Wiley-VCH Verlag GmbH & Co.

Subsequent publications explored the multifunctional metal carboxyphosphonates [41,43] via the method considered previously, successfully obtaining a range of structures based on Co and Mn. Over the years, several new MPs have been discovered thanks to the application of HT methods [45,49], whose efficacy was also proved for less conventional synthetic approaches, such as microwave-assisted synthesis [50] and ultrasonic synthesis [51].

One of the initial drawbacks identified for HT methods has been the inability to control the temperature for each individual reaction. Since temperature has shown to play such a profound role in determining the product outcome and structures obtained, it would be worth having more control of this factor, allowing for a larger parameter space than is already available to be explored. The solution to the problem of temperature control was found in a thermocycler, or a polymerase chain reaction machine, which is usually found in the field of genetics [39]. This has allowed for the control of temperature for individual reaction chambers.

3.2. Mechanochemical Synthesis

Mechanochemistry has a long history extending back over 1000 years [52,53] and had not, until recently, received mainstream attention as a viable tool in chemical synthesis. However, over the last two decades, there has been a massive push towards green and sustainable chemistry, which is perhaps one of the main reasons why mechanochemistry has since received a lot more attention in a range of specialised fields. While there are still a lot of details to be ironed out with regards to the specific mechanisms that drive mechanochemical synthesis, the basic principle is clear. The input of mechanical energy, i.e., through grinding/milling drives the reaction between two or more solids to form a desired product, which is often in the presence of little or no solvent [54]. One of the key points that makes mechanochemistry more appealing is that it is possible for reactions to proceed via pathways that are not accessible through conventional methods [55].

Some of the most significant work on the mechanochemical synthesis of MPs has been carried out over the past four years. In 2016, a synthesis was reported using a vibration ball mill, in which cadmium acetate dihydrate and phenylphosphonic acid were combined inside a reaction vessel in various ratios (1:1, 1:2, and 1:4) and then milled for 15 min along with two 4.0-g stainless steel balls. One known and two novel cadmium phenylphosphonates were obtained: Cd(O$_3$PPh)·H$_2$O (Figure 9a,d), Cd(HO$_3$PPh)$_2$ (Figure 9b,e), and Cd(HO$_3$PPh)$_2$(H$_2$O$_3$PPh) (Figure 9c,f) [56]. Each of the products obtained after milling were damp, which was caused by the release of water and acetic acid during the reaction, which in turn, means that each section of the synthesis was in fact liquid-assisted.

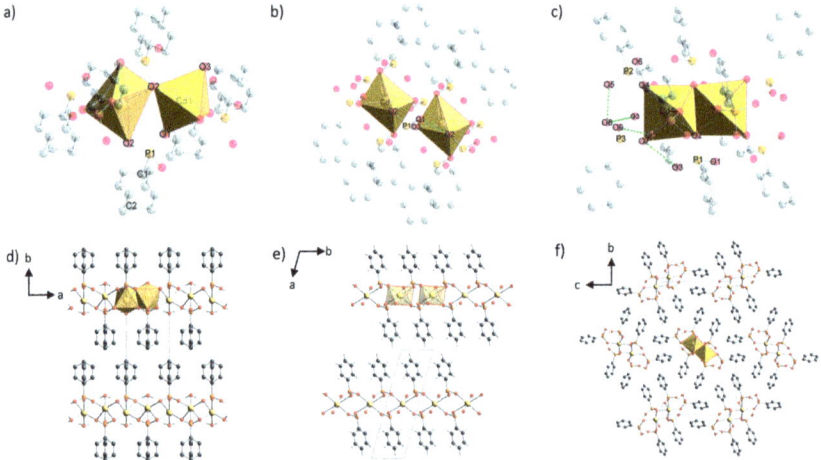

Figure 9. Molecular structure and coordination polyhedra for the Cd^{2+} ions and hydrogen bonds (green dashed lines) of (**a**) Cd(O3PPh)·H$_2$O, (**b**) Cd(HO$_3$PPh)$_2$, (**c**) Cd(HO$_3$PPh)$_2$(H$_2$O$_3$PPh), and the crystal structures of Cd(O$_3$PPh)·H$_2$O, (**d**) viewed along the *c*-axis, (**e**) Cd(HO$_3$PPh)$_2$, viewed along the *c*-axis, and (**f**) Cd(HO$_3$PPh)$_2$(H$_2$O$_3$PPh), viewed along the *a*-axis. Hydrogen atoms of the phenyl rings are omitted for clarity. Colour code: yellow: cadmium, orange: phosphorus, red: oxygen, grey: carbon, light grey: hydrogen. Adapted with permission from reference [56]. Copyright 2016, Royal Society of Chemistry.

In the same year, further work was carried out which looked at substituting cadmium with various other metals. The authors state that this is the first work that explores the mechanochemical synthesis of molecular MPs [57]. They present isomorphic structures of molecular MPs, M(HO$_3$PPh)$_2$(H$_2$O$_3$PPh)$_2$(H$_2$O)$_2$, where M = Mn, Co, and Ni (Figure 10a,b). Using the previously established procedure, the researchers were able to successfully synthesise three pure compounds after neat grinding for 15 min. They also carried out a liquid-assisted grinding (LAG) run for each of the compounds, finding that it had no effect on the data obtained when compared to the "dry" run. The conclusion was that this approach could be applied to the synthesis of other molecular MPs and would bring great benefits in terms of its simplicity and speed. There would also be opportunity for industrial applications due to the scalability, small environmental impact, and once again, the speed of synthesis.

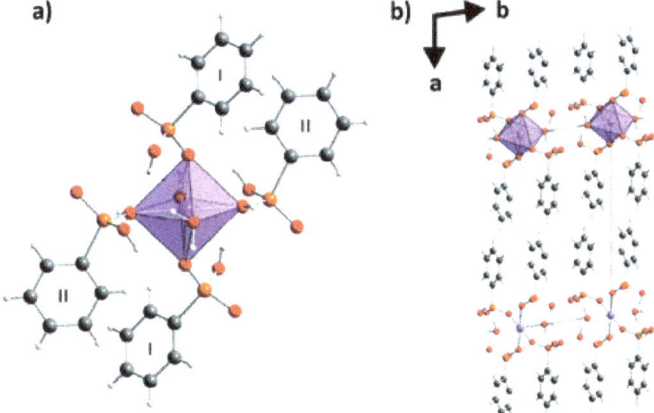

Figure 10. Crystal structure of [Mn(HO$_3$PPh)$_2$-(H$_2$O$_3$PPh)$_2$(H$_2$O)$_2$] (**1**), with respective values for the isomorphic compounds based on Co (**2**) and Ni (**3**). (**a**) Coordination sphere of the metal ion. (**b**) Structure of **1** shown along the *c*-axis (b). Hydrogen atoms are omitted for clarity. Colour code: purple: manganese, red: oxygen, orange: phosphorus, grey: carbon, light grey: hydrogen. Adapted with permission from reference. [57]. Copyright 2016, Royal Society of Chemistry.

More recent additions to the body of work looked at exploiting MPs with N-containing ligands. In 2017, the compound zinc N-(phosphonomethyl)glycinate Zn(O$_3$PCH$_2$NH$_2$CH$_2$CO$_2$)·H$_2$O was reported, displaying a 3D pillared-layered structure (Figure 11a–c) [58]. This was achieved through the LAG of zinc acetate dihydrate and N-(phosphonomethyl)glycine in a 1:1 ratio in the presence of water over 15 min. Through PXRD, it was shown that after just 30 s, the starting materials was almost completely consumed, and were completely absent after three minutes. Rietveld refinement proved that a pure compound is obtained. Further work on N-containing phosphonate ligands was carried out on manganese phosphonates [59], namely manganese mono(nitrilotrimethylphosphonate) (MnNP$_3$) and manganese bis(N-(carboxymethyl)-iminodi(methylenephosphonate)) [Mn(NP$_2$AH)$_2$].

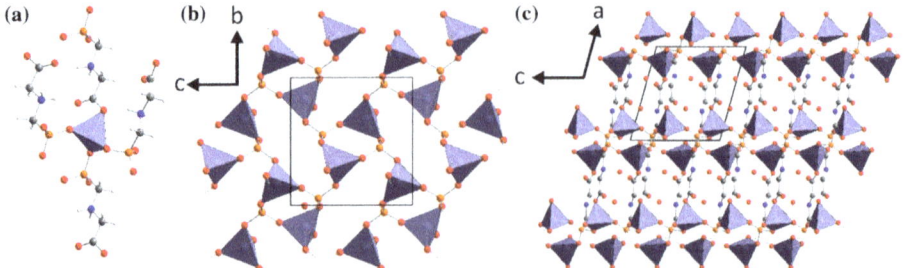

Figure 11. Structure of Zn($O_3PCH_2NH_2CH_2CO_2$)·H_2O. (**a**) Coordination sphere of the Zn^{2+} ion. (**b**) Structure of the layer formed by the ZnO_4-tetrahedra and the phosphono groups, shown along the *a*-axis. (**c**) Pillared structure of the framework shown along the *b*-axis (c). Hydrogen atoms are omitted for clarity. Colour code: Purple: zinc, red: oxygen, orange: phosphorus, blue: nitrogen, grey: carbon, and white: hydrogen. Adapted with permission from reference. [58]. Copyright 2017, Springer Science Business Media.

Numerous efforts have also been made to combine mechanochemical synthesis and in situ characterisation methods, which are discussed more in detail in Section 3.5, herein.

It is worth reiterating that mechanochemical synthesis provides a range of advantages over more conventional methods. It is a facile technique that provides results in relatively short periods of time, e.g., 15 min, and full conversion is often achieved. It is also worth considering the green status of the technique, owing to the requirement for little to no solvent. Then, it is clear that mechanochemical synthesis is an invaluable tool on any scale requirement, and could provide researchers with an alternative to conventional synthesis methods, yielding products that may have previously been inaccessible.

3.3. Porous Materials

The structures of most of the early MPs were two-dimensionally (2D) layered [15,16], whereby organic molecules would coordinate to a central metal layer via the phosphonate group. Then, the organic functionalities would interlace with other metal–organic layers, creating a dense, multi-layered system. As discussed in Section 2, some early examples of open framework MPs were reported in the 1990s which, with few exceptions [60], displayed pores too small to accommodate guest species larger than water, and whose porosity could not be probed by the adsorption of N_2 at 77 K [29]. In an almost parallel arc to open framework MPs, another class of porous inorganic–organic materials, i.e., metal–organic frameworks (MOFs), which are mainly based on carboxylic or N-heterocyclic ligands, started attracting growing interest from 1999 onwards [61–63]. Since then, MOFs have become one of the most intensely investigated classes of materials for applications such as gas separation and storage, catalysis, and drug delivery, owing to their large porosity, ease of functionalisation, and structural versatility [64,65]. Phosphonate-based MOFs represent a rather small fraction of the thousands of MOF structures reported over the last 20 years. This is mainly due to the strong tendency of MPs to form extended inorganic layers, coupled with the tendency of the phosphonate moiety to bridge metal ions. This poses significant synthetic challenges when the aim is to generate open frameworks. However, some interesting advances have been made recently in developing synthetic approaches that yield permanently porous MP materials [12–14,66].

One of the first examples of permanently porous MPs is MIL-91 [67]; this was based on the linker *N,N'*-piperazinediphosphonic acid, and was initially reported in 2006 by Serre et al. as two isoreticular analogues based on Al(III) and Ti(IV) (Figure 12a–c). The structures display monodimensional inorganic building units (IBUs) and channels running along the *b*-axis with an internal space of around 3.5 × 3.5 Å. A Langmuir surface area of around 500 $m^2·g^{-1}$, as well as a pore volume of 0.20 $cm^3·g^{-1}$, were measured by N_2 adsorption at 77 K. The compounds were also shown to be reasonably stable up to 550 K, after which there is a significant breakdown of the organic groups, causing the structure to become X-ray amorphous. The synthesis of both MIL-91(Al) and MIL-91(Ti) was successfully scaled up to the 100-g scale with milder procedures than those originally reported, i.e., using reflux instead of hydrothermal conditions, reducing the synthesis time and avoiding the use of hydrofluoric acid as a mineraliser [68]. A later study identified some unusual adsorption properties of the structure towards CO_2, which will be examined in detail in Section 4.2 [69].

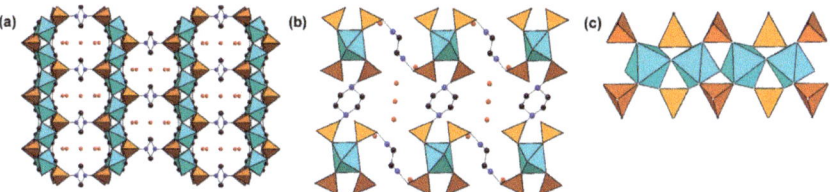

Figure 12. Crystal structure of MIL-91 viewed (**a**) along the *c*-axis and (**b**) the *b*-axis. For clarity, only one set of the two symmetry-equivalent (but half-occupied) sets of diphosphonate ligands connecting octahedral chains along *c* are shown. (**c**) A corner-sharing chain of MO_6 octahedra (c). Colour code: titanium: cyan, phosphorus: orange, oxygen: red, nitrogen: blue, carbon: purple. Adapted with permission from reference [67]. Copyright 2006, American Chemical Society.

The same ligand was also employed to prepare a series of isoreticular compounds (named STA-12) based on divalent metals, namely Ni, Co, Fe, and Mn, and featuring large hexagonal channels running along the *c*-axis (Figure 13) [70]. Interestingly, the STA-12 framework displays dehydration behaviour that is dependent on the metal atom: while STA-12(Ni) and STA-12(Co) are porous to N_2 when coordinated water molecules are removed from the metal ions in the IBUs, the Mn and Fe analogues are not [71]. This was attributed to the tilting of the piperazine rings and filling of the open coordination sites on Fe and Mn by the uncoordinated P–O group, which inhibits the uptake of N_2 probably by blocking access of the pores on the external surface. Using the linker *N*,*N*′-bispiperidinediphosphonic acid, an isoreticular analogue of STA-12, named STA-16, was obtained with Ni and Co as the metal atoms (Figure 13) [72]. Thanks to the longer linker, STA-16 features pores with 1.8-nm diameter, approaching the mesoporous regime, and a large pore volume of 0.68 cm^3·g^{-1}. So far, this remains a unique example of isoreticular expansion for porous MPs.

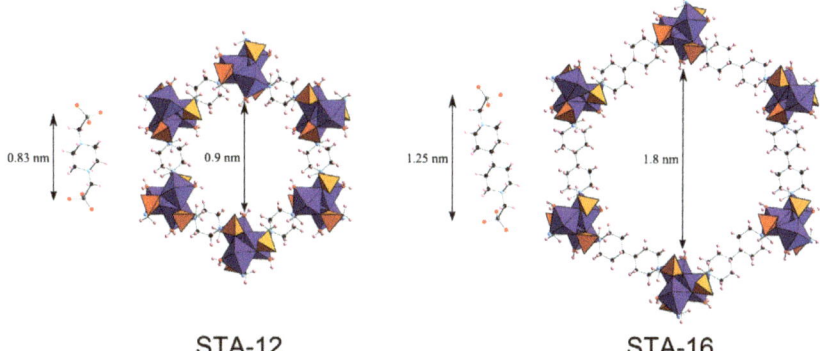

STA-12   STA-16

Figure 13. Crystal structures of (**left**) STA-12, based on *N*,*N*′-piperazinediphosphonic acid (**left**) and (**right**) STA-16, based on *N*,*N*′-bispiperidinediphosphonic acid (**right**). Colour code: cobalt, nickel: purple, phosphorus: orange, oxygen: red, nitrogen: light blue, carbon: black, hydrogen: white. Adapted with permission from reference [72]. Copyright 2011, American Chemical Society.

More recently, the use of rigid linkers with non-linear geometries was proven to be a viable strategy to induce the formation of porous structures [14]. Among them, tritopic linkers have been particularly investigated by several groups, and a recent review on the topic has been published, to which the interested reader is directed [73]. Besides tritopic linkers, tetratopic ones with either tetrahedral or square geometry have also proved successful in generating open framework compounds [74–77]. A notable example is the linker Ni-tetrakis(4-phosphonophenyl)porphyrin, which was combined with both divalent metals, namely, Mn, Co, Ni, and Cd [46,78], and tetravalent metals, namely Zr and Hf [47], to produce permanently porous compounds, termed CAU-29, CAU-36, and CAU-30, respectively (Figure 14a,b). HT methods were extensively employed to identify the best synthetic conditions to induce the formation of single crystals (in the case of divalent metals) and microcrystalline solids (in the case of tetravalent metals). Notably, the CAU-30 framework combines one of the most porous structures ever reported so far for MPs, featuring a specific surface area of almost 1000 $m^2 \cdot g^{-1}$ and a pore volume of 0.5 $cm^3 \cdot g^{-1}$, and exceptional stability, retaining its crystallinity up to 400 °C and pH 12.

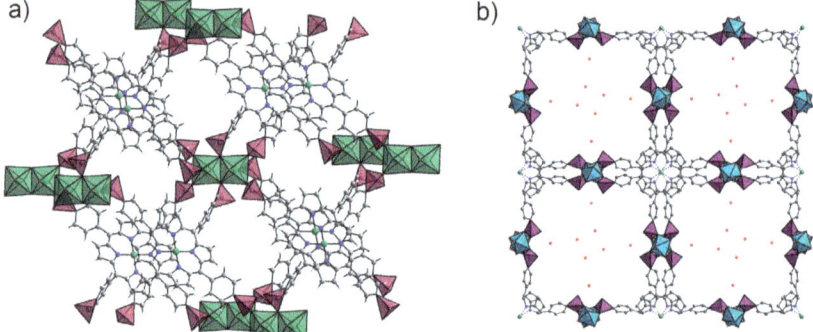

Figure 14. Crystal structures of (**a**) CAU-29 [Reprinted with permission from reference. [46]. Copyright 2018, Royal Society of Chemistry] and (**b**) CAU-30 [Reprinted with permission from reference [47]. Copyright 2018, Royal Society of Chemistry. Colour code: nickel: green, zirconium: light blue, phosphorus: purple, oxygen: red, nitrogen: dark blue, carbon: grey, hydrogen: white.

While the field of carboxylate-based MOFs has moved beyond the initial phase—which was dominated by crystal engineering—towards practical applications and large-scale deployment, progress in phosphonate-based MOFs has been slower, and it still remains challenging to identify the right combination of linker, metal, and synthetic conditions to generate permanently porous structures. This section has showed that the recently renewed interest in microporous MPs has already produced some notable results, which hold promise for further advances in the near future.

3.4. Phosphinic Acid-Based Materials

All previous discussion in this review has focused on phosphonic acid-based materials. However, a closely related class of compounds has also been explored: so-called phosphinic acid-based materials. Phosphinic acids are different from phosphonic acids, in that one of the –OH groups on phosphorous is replaced by a second organic moiety. This makes phosphinates less versatile in terms of coordination ability, but at the same time provides the possibility to exploit the nature of the organic moieties attached to the P atom. A comparison of the structure of phosphonic and phosphinic acids can be seen in Figure 15. Thus far, metal phosphinates have received less attention than phosphonates, and most of the reported structures are based on either diphenylphosphinic acid or diphosphinates with small organic groups, which are sometimes combined with auxiliary ligands [79,80]. Besides their similarity with phosphonic acids, phosphinic acids also share some features with carboxylic acids: they are monovalent groups with two oxygen atoms available to coordinate to metal species, although the different hybridisations of the phosphinic P atom (sp^3) and the carboxylic C atom (sp^2) have obvious effects on the geometry of the two groups (Figure 15). Notably, phosphinic acids are more acidic than carboxylic acids, which is the reason why metal phosphinates have shown much better hydrolytic stability. As mentioned above, carboxylic acids have usually been the linkers of choice in "traditional" MOF chemistry. A few examples of microporous compounds based on phosphonate monoesters (Figure 15), which share some features with phosphinic acid linkers, have been reported recently [81–84]. On the other hand, until very recently, no permanently porous metal phosphinate MOFs were reported.

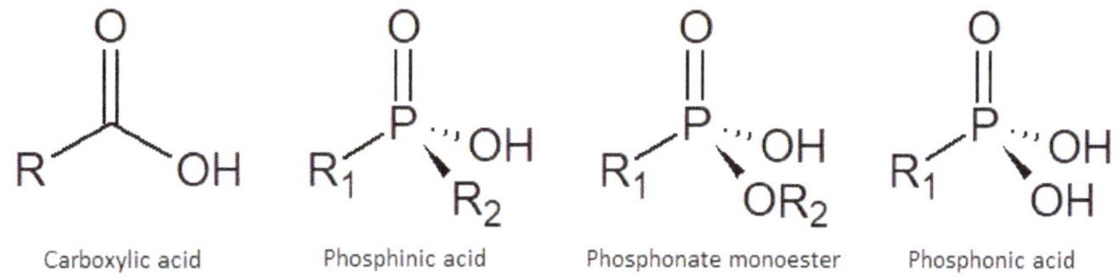

Figure 15. Molecular structures of carboxylic, phosphinic, and phosphonic acids.

In 2018, Hynek et al. [82] used the linker phenylene-1,4-bis(methylphosphinic acid), hereafter denoted PBPA(Me), which can be seen as an analogue of terephthalic acid, the most common linker in MOF chemistry. Three different compounds were obtained by the reaction of PBPA(Me) with Fe(III) salts in different conditions (Figure 16): ICR-1 (ICR stands for Institute of Inorganic Chemistry Řež), prepared in water, has a non-porous 3D framework, which is isoreticular to a previously reported Zr phosphonate [85]; ICR-2, prepared in ethanol, is instead porous, with hexagonal channels with diameters of 0.8 nm running along the *c*-axis and a Brunauer-Emmett-Teller (BET) surface area of 731 m^2·g^{-1}; ICR-3, prepared in N,N'-dimethylformamide is a layered dense phase. ICR-2 was shown to be stable in a range of solvents, most notably water. The authors were also able to obtain a compound isoreticular to ICR-2, termed ICR-4, by employing the linker phenylene-1,4-bis(phenylphosphinic acid) [PBPA(Ph)]. Due to the larger size of the organic groups accommodated in the pores, ICR-4 has a much lower surface area (165 m^2·g^{-1}) than ICR-2.

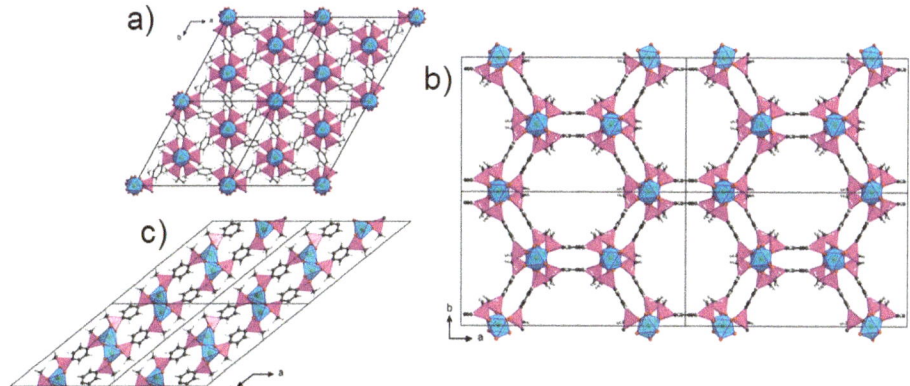

Figure 16. Crystal structures of (a) ICR-1, (b) ICR-2, and (c) ICR-3. Colour code: iron: blue, phosphorus: pink, oxygen: red, carbon: grey, hydrogen: white. Adapted with permission from reference. [82]. Copyright 2018, Wiley-VCH Verlag GmbH & Co.

The results reported by Hynek et al. demonstrate that phosphinic acids can also be suitable linkers for the construction of open framework structures, with the added advantage that the pendant organic group can be modified and/or functionalised to impart specific properties to the material. This is not possible in the case of phosphonates and carboxylates; therefore, the use of phosphinic acids might open up a whole new range of opportunities in the field of MOFs.

3.5. Structure Determination by Electron Diffraction

2018 has seen a major breakthrough for the structural characterisation of microcrystalline and nanocrystalline compounds, mostly thanks to two major papers reporting on the use of electron diffraction (ED) data to solve the crystal structures of some small organic molecules [86,87]. Using ED, it is possible to obtain diffraction patterns that are comparable to those usually observed when performing SCXRD experiments from crystals of submicrometric size. The method is particularly attractive because it does not require exotic instrumentation and, in principle, any transmission electron microscope could be adapted for such a task, holding promise for a rapid deployment of the technique as a routine tool for structural characterisation.

Even before the two aforementioned papers appeared, ED had already been employed to determine the crystal structure of some MOFs [88], including some compounds mentioned earlier in this review: CAU-30, CAU-36, (Section 3.3) [47,78], ICR-1, ICR-2, and ICR-3 (Section 3.4) [82]. All these compounds were obtained as microcrystalline powders, whose crystallite size was below the micrometre range and not sufficiently high to allow structure solution from PXRD data (Figure 17). In the case of CAU-30, the structural model was obtained from automated electron diffraction tomography [89], whereas for the ICR compounds, micro ED [90] was employed. In order to minimise beam damage, the samples were cooled to 95 K (CAU-30) or 100 K (ICRs), and the beam was shifted on the crystals. For all the compounds, the structural model was then validated by classical Rietveld refinement of the PXRD patterns. Another notable example of the use of ED in MP chemistry is represented by CAU-36 [78], which was a Co-based porous compound obtained using the same Ni-tetrakis(4-phosphonophenyl)porphyrin linker employed to prepare CAU-29 and CAU-30. In the case of CAU-29, continuous rotation electron diffraction [91] was employed, allowing to precisely identify the location of the guest species (1,4-diazabicyclo[2.2.2]octane and solvent molecules) within the framework.

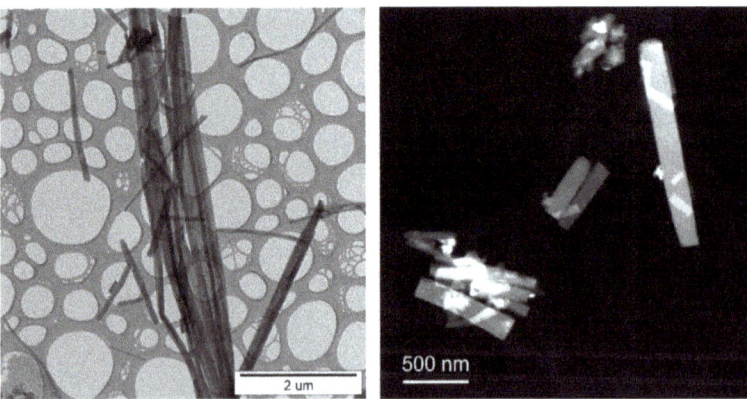

Figure 17. Transmission electron micrographs of ICR-2 (**left**) (Reprinted with permission from reference. [82]. Copyright 2018, Wiley-VCH Verlag GmbH and Co.) and CAU-30 (**right**) (Reprinted with permission from reference. [47]. Copyright 2018, Royal Society of Chemistry).

Given the difficulty in obtaining suitable crystals for SCXRD, especially when high oxidation state metals are employed, the structural determination of MPs has often been carried out from PXRD data. However, due to the limited amount of information contained in a PXRD pattern, some structures can require a large amount of time to be solved, if they can be solved at all. This is one of the main factors that have prevented the field of MPs from progressing at the pace of other classes of coordination polymers, such as carboxylate-based MOFs. Therefore, the possibility of accessing ED as a tool for structural determination can be an absolute game changer in this sense.

3.6. In Situ Characterisation

In situ investigations have been widely used in materials chemistry to monitor synthetic processes, especially those involving the self-assembly of well-defined building blocks into crystalline compounds [92]. Besides simply following the products' formation, this approach is powerful in identifying the different phases and intermediates that form over the course of a reaction [93]. The application of in situ investigations can take many different forms and draw on a wide range of techniques, such as XRD, X-ray absorption spectroscopy, and vibrational spectroscopies [94]. The data obtained from in situ studies can be used to identify the different stages in a reaction and correlate this to the reaction parameters for further optimisation.

An early example of in situ investigations of MPs stems from Feyand at al. and the work carried out on the HT synthesis of lanthanide phosphonatobutanesulfonates, $Ln(O_3P-C_4H_8-SO_3)(H_2O)$, where Ln = La – Gd [95]. They managed to produce a series of isostructural compounds on which they carried out in situ analysis using energy-dispersive X-ray diffraction (EDXRD) during both conventional and microwave-assisted heating processes. A number of phases were identified before the formation of the final product (Figure 18). During the first five minutes, there was some background modulation, indicating that an amorphous side phase had formed in the initial part of the reaction, but after one minute, there was clear evidence for a crystalline intermediate. The formation of the product was observed at seven minutes, after which no significant events were identified in the diffraction pattern. The researchers concluded that the reaction takes place in two steps between 110–150 °C for both conventional and microwave heating, with no observable difference in the phase change, but some in the ongoing crystallisation of the product.

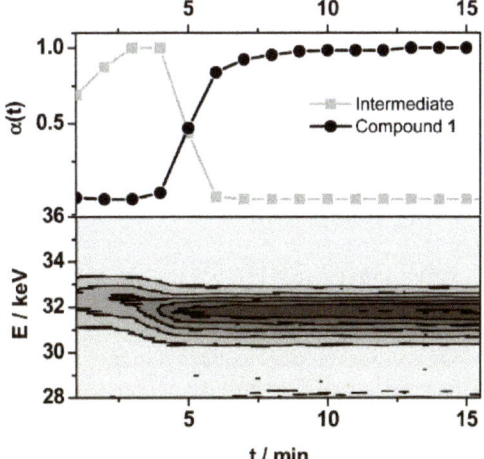

Figure 18. Surface energy-dispersive x-ray diffraction (EDXRD) plot of the transformation of the intermediate phase into $Ln(O_3P-C_4H_8-SO_3)(H_2O)$ (bottom) and reaction progress ($\alpha(t)$) for both phases under conventional heating at 150 °C (top). Reprinted with permission from reference. [95]. Copyright 2010, American Chemical Society.

The same group also carried out a temperature-dependent in situ EDXRD study for the formation and temperature-induced phase transition of previously described copper phosphonatosulfonates [96]. It was shown that at 90 °C, the formation of $[Cu_2(O_3PC_2H_4-SO_3)(OH)(H_2O)]\cdot 3H_2O$ is achieved after five minutes, but proceeds via a previously unknown phase, i.e., a tetrahydrate analogue of the previous compound, which they were able to obtain as a phase pure product by quenching the reaction mixture within the first few minutes. Alternatively, increasing the temperature from 90 °C to 150 °C saw the formation of the monohydrate analogue after 12 min, with full phase transformation at 15 minu. This in situ EDXRD investigation allowed the researchers to propose the likely reaction pathway, which was shown in Figure 19.

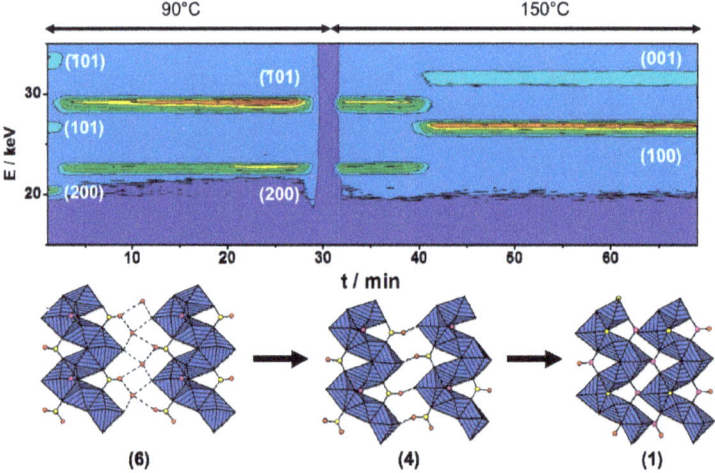

Figure 19. Illustration of the reaction pathway leading to the formation of $[Cu_2(O_3PC_2H_4\text{-}SO_3)(OH)(H_2O)]\cdot 3H_2O$ **(1)** through intermediates $Cu_2[(O_3P\text{-}C_2H_4\text{-}SO_3)(H_2O)_2(OH)]\cdot 4H_2O$ **(4)** and $[Cu_2(O_3P\text{-}C_2H_4\text{-}SO_3)(H_2O)_2(OH)]\cdot 3H_2O$ **(6)**. The phase transformation at different temperatures is shown in the contour plot (top), and the correlated structural motifs of the copper oxygen chains are presented at the bottom. The gap in the contour plot at ~30 min is due to the stage at which the specimen was removed from the vessel. Reprinted with permission from reference. [96]. Copyright 2012, American Chemical Society.

In 2017, Heidenreich et al. [93] published a detailed overview of a novel reaction cell with integrated stirring and heating systems (up to 200 °C), which was named *SynRAC*, and used for in situ investigation of reactions under hydrothermal or solvothermal conditions using synchrotron radiation. The setup and design of the cell is discussed in detail in the paper: the main advantage of the *SynRAC* is that it is designed to be as similar as possible to a common laboratory reaction vessel, which allows for the extensive and reliable preliminary ex situ screening of reaction parameters before the actual in situ work, thus helping maximise the amount of information obtainable from the available synchrotron beamtime. A number of case studies are also presented to demonstrate the methods they use and the results that can be achieved. These do not include any MPs, but the *SynRAC* could easily find employment for this class of materials as well.

Section 3.2 has already suggested that in situ studies can be very informative for mechanochemical syntheses [56–59,97]. The two techniques used in each of the in situ studies reported so far were XRD and Raman spectroscopy, which allow the reaction to be monitored at both molecular and crystalline levels [97]. Traditionally, the vessel used in the milling/grinding process would be lined with an abrasion-resistant material, and would contain the grinding media, which is often constructed from steel, ceramic, and other suitable materials. However, for the in situ studies, the vessel needed to be adapted, which led to the grinding vessel to be made of Perspex. This allows radiation from both techniques to penetrate the vessel and reach the appropriate detector.

Combined XRD and Raman data on metal phenylphosphonates [57] has allowed the researchers to identify various phases throughout reactions. In 2015, Batzdorf et al. [97] described the in situ investigation for the synthesis of cobalt(II) phenylphosphonates monohydrate (CoPhPO$_3$·H$_2$O), where equimolar amounts of cobalt(II) acetate and phenylphosphonic acid were ground together to form the product. The in situ XRD and Raman data confirmed that the product started forming after only 2.75 min of grinding, with completion of the reaction occurring at 27 min (Figure 20a,b). Different stages of the reaction were also identified, the main being where the product coexisted with intermediate phases.

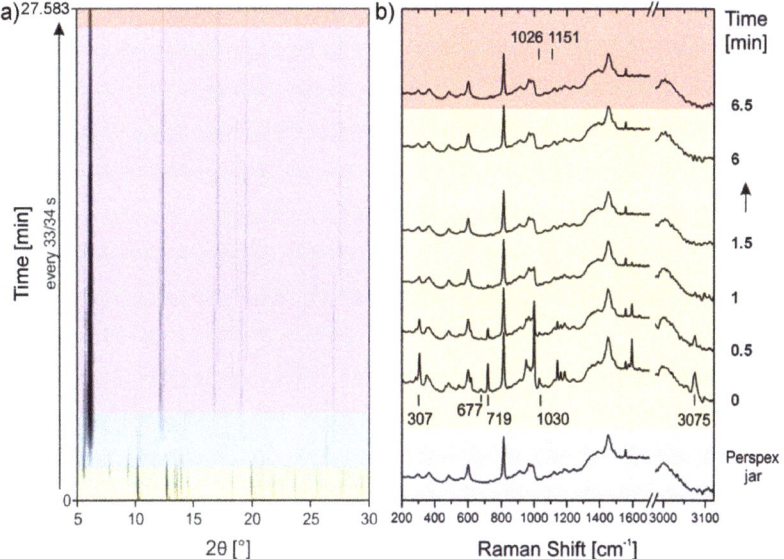

Figure 20. Synthesis process of Co(PhPO$_3$)·H$_2$O followed in situ by (**a**) synchrotron X-ray diffraction (XRD) and (**b**) Raman spectroscopy. The first measurement of the Raman plot (grey) is the Raman spectrum of the empty Perspex jar, indicating the modes arising from the sample holder. Colour code: yellow: reactants, green: reactants and intermediate phases, blue: reactants, intermediate phases and product, purple: intermediate phases and product, red: product. Adapted with permission from reference [97]. Copyright 2015, Wiley-VCH Verlag GmbH and Co.

An example where in situ analysis was able to provide unparalleled insight is the formation of Mn(HO$_3$PPh)$_2$(H$_2$OPPh)$_2$(H$_2$O)$_2$. XRD data indicated that there were five phases during the reaction [57] (Figure 21a,b). The initial reflections represent the starting materials (phase 1), and it is not until 30 s into the reaction that additional reflections indicate the appearance of a new phase (phase 2), which itself lasts approximately 30 s. At this point, strong reflections for the final product can be observed alongside those of Mn(O$_3$PC$_6$H$_5$)·H$_2$O (phase 3). The reflections for both starting materials disappear at 1.15 min, after which the Mn(O$_3$PC$_6$H$_5$)·H$_2$O reflections start to decrease, while those of the final product intensify (phase 4). Then, at six minutes, the reflections for anything other than the product reach a minimum and go through no further changes (phase 5). Raman data, after 30 s of milling, shows bands exclusively for the uncoordinated phenylphosphonic acid. After an additional 30 s, bands assigned to the coordinated phenylphosphonic acid start to appear, followed by the disappearance of the uncoordinated phosphonic acid band. The final change is the increase in intensity for the coordinated phenylphosphonic acid band.

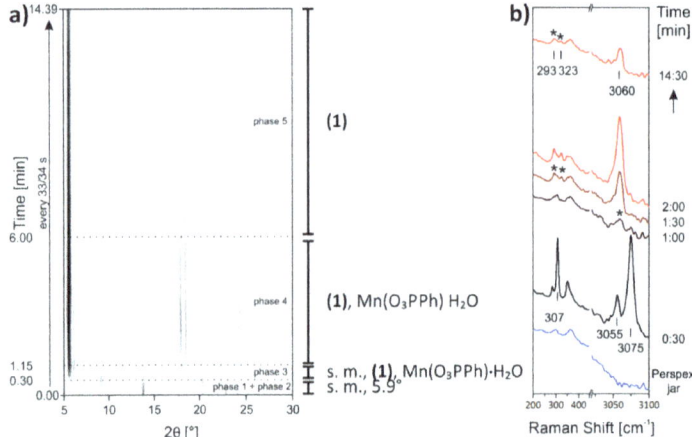

Figure 21. Two-dimensional (2D) plots of (**a**) synchrotron X-ray diffraction (XRD) data with a description of the detected compounds and (**b**) in situ Raman spectroscopy measurements monitoring the synthesis of Mn(HO$_3$PPh)$_2$(H$_2$O$_3$PPh)$_2$(H$_2$O)$_2$. Adapted with permission from reference. [57]. Copyright 2016, Royal Society of Chemistry.

Based on the aforementioned examples, it is clear that the use of in situ investigations can provide invaluable information on the mechanism of a reaction as well as the different crystalline phases appearing before the reaction completion. Furthermore, they present opportunities to optimise conditions to isolate specific new materials that might exist as intermediate phases for products with previously known structures.

4. Applications of Metal Phosphonates

4.1. Catalysis

Catalysis has played a key role in enabling the huge progress of the chemical industry in the 20th century [98]. Most of the large-scale industrial processes producing essential commodities, such as ammonia, sulfuric acid, nitric acid, and polyolefins, are made possible by the use of heterogeneous catalysts [99,100]. Typically employed catalysts in industrial settings consist of metal nanoparticles deposited on the surface of solid supports, such as metal oxides, carbon-based materials or zeolites [101,102]. With the advent of MOFs and the precise atomic control that they allow, it has also become possible to introduce active sites, which are most often coordinatively unsaturated metal atoms, in the porous framework [103,104]. Drawing inspiration from earlier work on the use of zirconium and titanium phosphates as solid acid catalysts [105], initial applications of MPs in catalysis were focused on, taking advantage of the Brønsted acidity of pendant sulfonic groups, which were successfully employed for cracking reactions in mild conditions [106]. This section presents recent examples of MPs employed both as supports for catalytically active species and as catalysts themselves.

A crucial feature of effective catalyst supports is a large surface to make the active species easily accessible to reactants. This can be achieved by the use of porous supports or by the exfoliation of layered materials to generate nanosheets consisting of single or few layers. A representative example of the latter strategy involves a mixed zirconium phosphate/phosphonate of formula Zr$_2$(PO$_4$)H$_5$[(O$_3$PCH$_2$)$_2$NCH$_2$COO]$_2$·H$_2$O (Figure 22) containing N,N-bis[(phosphonomethyl)glycine] (glyphosine) as a ligand [107]. The interlayer region of this structure is characterised by the presence of both carboxylic and phosphonic acid groups, which make the layer surface highly polar and give rise to an extended network of hydrogen bonds. The presence of these groups make the compound very prone to ion exchange and the intercalation of small organic amines, such as propylamine, yielding stable colloidal dispersions of hybrid nanosheets. The addition of palladium acetate to the dispersion led to the coordination of the metal to the non-coordinating carboxylic and phosphonic acid groups and enabled the deposition of palladium nanoparticles with size <2 nm onto the nanosheets with variable loadings, with the highest being reported at 19 wt. % [108]. The catalyst loaded with 15 wt. % of Pd was tested for the Suzuki–Miyaura coupling between phenylboronic acid and various aryl bromides, both in batch and continuous flow conditions, showing excellent performance and recyclability (Table 1 and Scheme 1). Importantly, the minimal leaching of Pd and no significant nanoparticle sintering were observed, suggesting the effectiveness of the support in binding the catalytically active species. The same catalyst was later successfully employed also for the hydrogenation in batch conditions of alkynes and nitroarenes [109] and for the Heck reaction [110], proving its wide applicability.

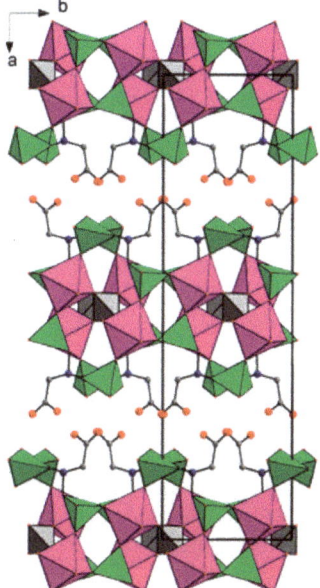

Figure 22. Crystal structure of $Zr_2(PO_4)H_5[(O_3PCH_2)_2NCH_2COO]_2 \cdot H_2O$ viewed along the *c*-axis. Colour code: zirconium: pink, phosphorus: green, oxygen: red, nitrogen: blue, carbon: grey. Reprinted with permission from reference. [107]. Copyright 2014, American Chemical Society.

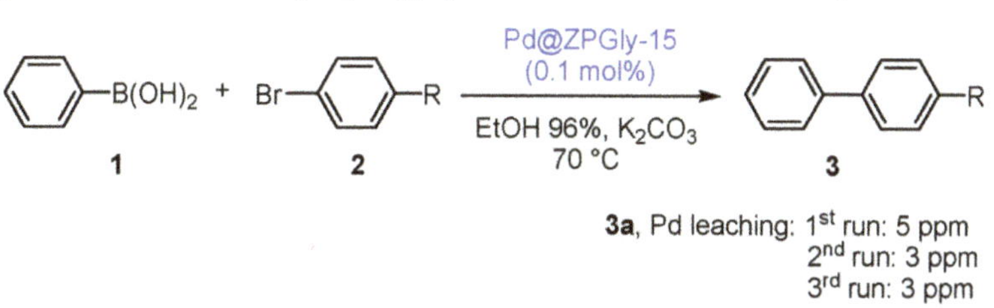

Scheme 1. Representative Suzuki–Miyaura coupling catalysed by Pd nanoparticles supported on $Zr_2(PO_4)H_5(O_3PCH_2)_2NCH_2COO)_2 \cdot H_2O$ nanosheets (15 wt. % Pd). Reprinted with permission from ref. [108]. Copyright 2015, Royal Society of Chemistry.

Table 1. Results of catalytic tests for the Suzuki–Miyaura coupling displayed in Scheme 1. Reprinted with permission from ref. [108]. Copyright 2015, Royal Society of Chemistry.

Entry	Aryl Bromide	R	t (h)	Yield (%)
1	2a (1st)	Me	30	97
2	2a (2nd)	Me	30	97
3	2a (3rd)	Me	30	97
4	2b	H	30	96
5	2c	CHO	10	98
6	2d	NO_2	10	98

More recently, a new strategy was adopted to obtain a zirconium phosphate–phosphonate decorated with very small gold nanoparticles on its surface [111]. The procedure consisted of different steps: first, a nanosized zirconium phosphate was obtained using a synthesis from gel in propanol [112]. Then, due to the small dimensions and high defectivity of the nanocrystals, this compound could be functionalised with amino groups through a topotactic anion exchange reaction of phosphates with incoming aminoethylphosphonate groups, and a compound with formula $Zr(PO_4)_{1.28}(O_3PC_2H_4NH_2)_{0.72}$ (hereafter ZP-AEP) was prepared (Figure 23). Finally, gold nanoparticles with dimensions less than 10 nm were produced by reduction, with $NaBH_4$, of $AuCl^{4-}$ ions chemisobted on the surface of nanocrystals (Figure 24).

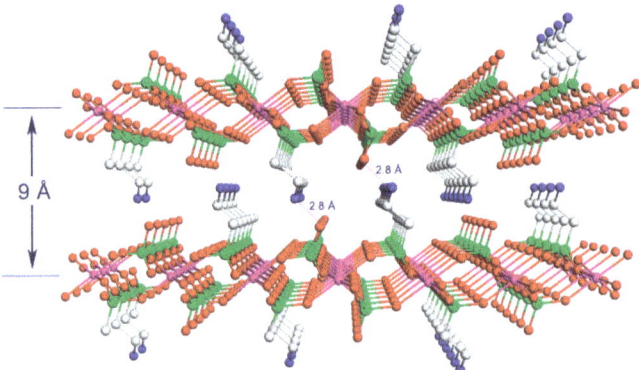

Figure 23. Schematic representation of a mixed α-Zr phosphate aminoethylphosphonate. Adapted with permission from reference [111]. Copyright 2019, Royal Society of Chemistry.

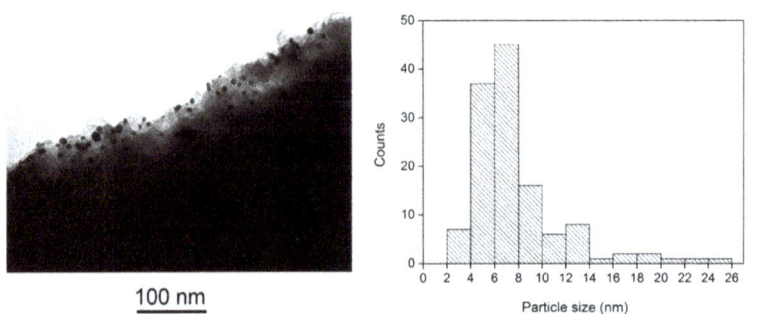

Figure 24. Transmission electron micrograph of Au@ZP-AEP (**left**) and Au particle size distribution (**right**). Adapted with permission from reference [111]. Copyright 2019, Royal Society of Chemistry.

A sample obtained with this procedure, and containing 1 wt. % Au, was found to be an efficient catalyst for the chemoselective reduction of a series of nitroarenes (Scheme 2); furthermore, the authors showed that it was possible to efficiently switch the reaction product between the corresponding azoxyarenes or anilines, simply by changing the solvent (96% EtOH or absolute EtOH, respectively). The recovery and reuse of the catalyst was also very efficient, as shown in Table 2.

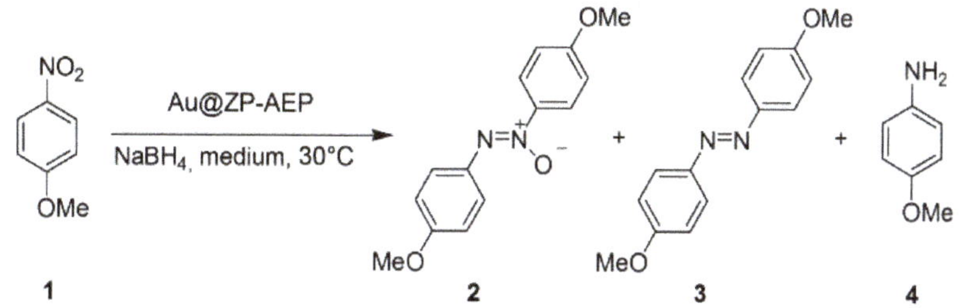

Scheme 2. Representative reduction of nitroarenes using Au@ZP–AEP as the catalyst and NaBH$_4$ as the reducing agent.

Table 2. Representative results of Au@ZP–AEP catalyst in the switchable preparation of an azoxy derivative (**2a**) and methoxyaniline (**4a**) and its recovery and reuse. Reprinted with permission from reference. [111]. Copyright 2019, Royal Society of Chemistry.

Entry [a]	Run	Medium	t (h)	C [b] (%)	2 : 3 : 4 Ratio [b]
1	Run 1	96% EtOH	3	98	96 : 4 : 0
2	Run 2	96% EtOH	3	96	97 : 3 : 0
3	Run 3	96% EtOH	3	94	97 : 3 : 0
4	Run 4	96% EtOH	3	93	96 : 4 : 0
5	Run 5	96% EtOH	3	87	98 : 2 : 0
6	Run 1	EtOH$_{abs}$	2	>99	0 : 0 : 100
7	Run 2	EtOH$_{abs}$	2	>99	0 : 0 : 100
8	Run 3	EtOH$_{abs}$	2	>99	0 : 0 : 100
9	Run 4	EtOH$_{abs}$	2	>99	0 : 0 : 100
10	Run 5	EtOH$_{abs}$	2	>99	0 : 0 : 100

[a] Reaction conditions: **1** (0.1 mmol), Au@ZP–AEP (1 mol%), NaBH$_4$ (6 equivalents), reaction medium: 96% EtOH or EtOH$_{abs}$ (1.8 mL), at 30 °C. [b] Conversion and product ratios determined by gas liquid chromatography and ^{1}H-NMR analyses.

A notable feature of both STA-12(Ni) and STA-12(Co) is their ability to be activated in order to create coordinatively unsaturated metal sites [113]. This occurs due to the loss of chemisorbed water at around 85–109 °C. Given the high density of these Lewis acidic metal sites, the STA-12 family of MPs has been investigated for a range of catalytic reactions. STA-12(Co) has been particularly notable for the catalysis of aerobic epoxidation of olefins, which are typically catalysed by Co-doped zeolites [114]. The issue with zeolitic materials in this case is that in order to achieve the required isolated Co species, the loading of Co in the zeolite must be relatively low, which means that larger quantities of the catalyst will be required. This is where MOFs have shown promise over other materials, in that they are often, by nature, high metal-containing materials. The catalytic activity of STA-12(Co) was found to be comparable to the zeolite catalysts, obtaining similar results even when reducing the amount of catalyst by two orders of magnitude. It was also found that the substrate used could have a great effect on the selectivity. With styrene, the selectivity to styrene oxide was low due to substrate oligomerization. Conversely, for (E)-stilbene, and (Z)-stilbene, the selectivities for the relative oxides were shown to be between 80–90%, with no considerable oligomerization of the substrate (Figure 25). It was also shown that the catalyst could be recycled, with little compromise on the activity. STA-12(Ni), similar to its Co analogue, has also shown reasonable catalytic activity. Both materials also showed little or no change in crystallinity when reused, with only a 1% difference in product formation after three cycles.

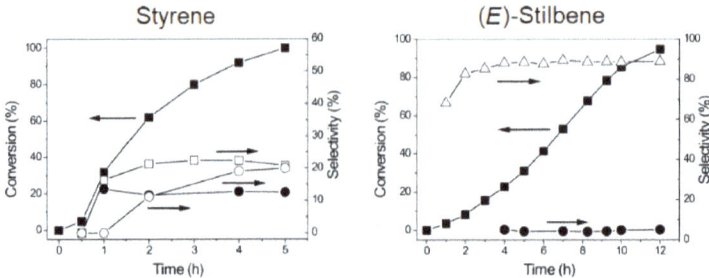

Figure 25. The epoxidation reaction of styrene (**left**) and (*E*)-stilbene (**right**) with the Co-based metal–organic framework material STA-12(Co). The graph on the left shows styrene conversion (■), as well as the selectivity for styrene oxide (□), benzaldehyde (●), and benzoic acid (○). The graph on the right shows (*E*)-stilbene conversion (■), as well as the selectivity for *trans*-stilbene oxide (△) and benzaldehyde (●). Adapted with permission from reference. [114]. Copyright 2012, Wiley-VCH Verlag GmbH & Co.

The combination of the high stability and structural versatility of MPs can be features promoting their employment in heterogeneous catalysis. As for MOFs, the presence of organic groups prevents their employment in high-temperature processes similar to those that are typical of the heavy petrochemical industry. However, they could potentially find application in the production of fine chemicals and catalytic processes of importance for sustainability issues, such as the conversion of CO_2 into value-added feedstocks, methane oxidation to methanol, electrocatalytic or photocatalytic hydrogen production, and water oxidation, which involve less harsh conditions.

4.2. Gas Sorption and Separation

Microporous MPs have the potential to be ideal candidates for gas sorption due to their high stability, especially in humid conditions, which often exceeds that displayed by carboxylate-based MOFs [12]. Unfortunately, as already discussed in Section 3.3, it has proved difficult for researchers to obtain porous MPs, since they are prone to forming densely packed layered structures, and only a small number have been reported to date [14,73]. Microporous MPs usually feature relatively low surface areas and pore volumes, which limits their scope for gas storage. However, most of them display channel-like pores, with diameters often below 10 Å, which can in principle provide favourable interaction between specific adsorbate species and the sorbent surface, which is a key requirement for the efficient separation of gaseous mixtures [115,116].

One of the most promising porous MPs identified for gas separation so far is MIL-91(Al/Ti), which features channels 3.5 × 3.5 Å in size [67,117]. On carrying out the adsorption experiments on CO_2, CH_4, and N_2 at 303 K, a strong affinity of MIL-91(Al) for CO_2 was observed (Figure 26) [69]. For both CH_4 and N_2, the isotherms showed no significant uptake of the gases, and failed to show any saturation plateau, even at ~50 bar. On the other hand, the isotherm for CO_2 showed significant uptake, with the majority of the gas being adsorbed at pressure below 1 bar, and complete saturation reached at ~15 bar. Microcalorimetric experiments showed that the enthalpy of CO_2 adsorption is constant at about 40 kJ·mol^{-1} up to a loading of 4 mmol·g^{-1}, suggesting a strong physisorptive character, which is likely induced by the close interaction of CO_2 with the surrounding pore walls.

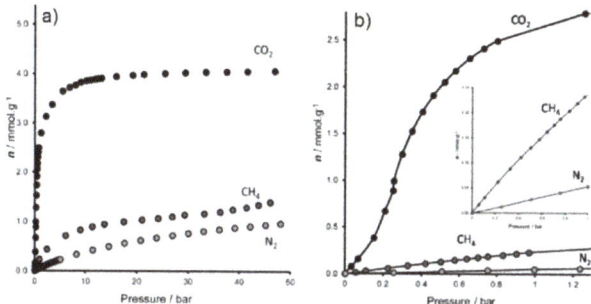

Figure 26. CO_2, CH_4, and N_2 adsorption isotherms measured at 303 K on MIL-91(Al) up to 50 bar (**left**) and up to 1 bar (**right**). Adapted with permission from reference. [69]. Copyright 2015, American Chemical Society.

Furthermore, an inflection on the CO_2 adsorption isotherm was identified at low pressure (Figure 26), which is due to a cooperative phenomenon involving the twisting of the ligand when a threshold CO_2 pressure is reached (Figure 27). This "phase-change" behaviour is especially interesting for its potential to combine the high working capacity and low energy penalty for the regeneration of the sorbent. The MIL-91(Ti) analogue showed a similar preference for CO_2 over the other gases, with a steep uptake at low pressure and saturation exceeding 4.0 mmol·g^{-1} [68]. However, unlike the Al analogue, no flexibility in the structure was observed, which was exemplified by the lack of inflection or S-character in the isotherm. Overall, MIL-91(Ti) was identified to be a viable material for CO_2 capture, owing to increased thermal stability when compared to other MOFs, its ability to selectively adsorb CO_2 over other gases, and the possibility to be produced in quantities beyond the laboratory scale.

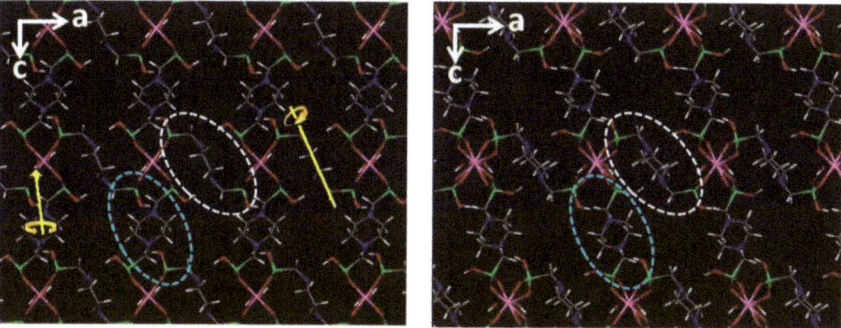

Figure 27. Crystal structure of MIL-91(Al) before (**left**) and after (**right**) the adsorption of CO_2. The rotation of crystallographically independent linker units is highlighted within the white and blue circles. Reprinted with permission from reference. [69]. Copyright 2015, American Chemical Society.

The gas separation properties of the STA-12 framework were also thoroughly investigated for the analogues containing Ni, Co, and Mg [71]. IR experiments with CO and CO_2 as probes showed that despite the presence of coordinatively unsaturated metal sites, none of the STA-12 frameworks display a strong interaction of the adsorbates with these sites. By the adsorption of N_2 at 77 K and CO_2 at 195 K, it was found that STA-12(Ni) is the most porous compound of the series, and further characterisation demonstrated that it has a high selectivity for CO_2 over CH_4 at ambient temperature [118]. The separation performance of STA-12-Ni was compared with that of the CPO-27-M (M = Ni, Co, and Zn), which is a carboxylate-based MOF with similar pore structure and a high density of open metal sites exposed on the channels [119]. The MOFs were tested for both binary

50:50 CO_2/CH_4 mixtures and ternary 70:15:15 $CO_2/CO/CH_4$ mixtures by means of single gas isotherms and breakthrough analysis (Figure 28a–d). The CPO-27 frameworks outperformed STA-12 for the separation of the binary mixture, thanks to the strong interaction of CO_2 with the open metal sites. On the other hand, when exposed to the ternary mixture, CPO-27-Ni and CPO-27-Co displayed preferential adsorption of CO, whereas STA-12(Ni) and CPO-27-Zn maintained their selectivity for CO_2. STA-12(Ni) crystallises as submicron-size particles suitable for use as the stationary phase in a porous layer open tubular (PLOT) capillary column, which was used to separate alkanes according to boiling point, giving promising separation performance, even without optimization [71]. The stability and versatility of MOF structures suggest that they can find specialist application in this field.

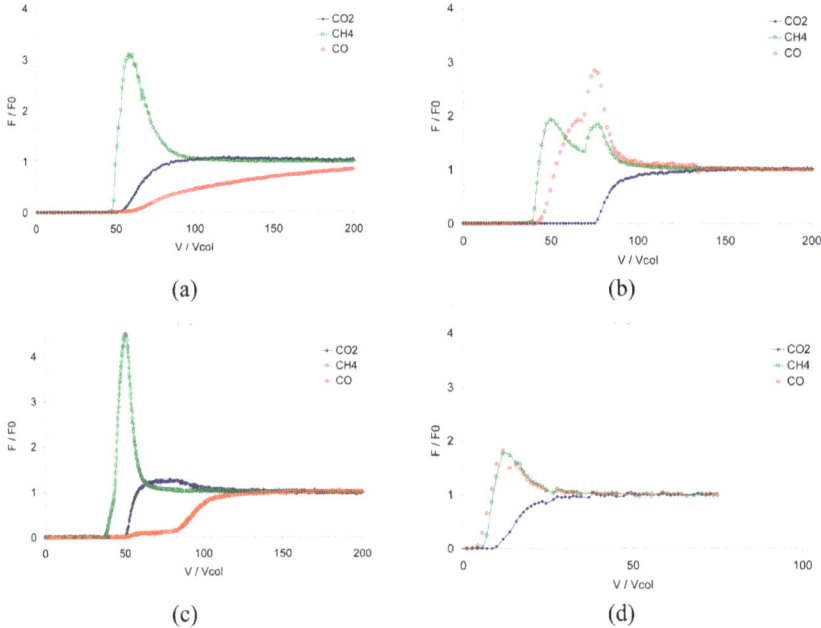

Figure 28. Breakthrough curves of the $CO_2/CH_4/CO$ (70/15/15) mixture on (**a**) CPO-27-Co, (**b**) CPO-27-Zn, (**c**) CPO-27-Ni, and (**d**) STA-12-Ni at 303 K and 5 bar. Reprinted with permission from reference. [71]. Copyright 2011, Elsevier Inc.

Given the recent progresses made in synthesising porous MPs, more intensive investigation of their gas sorption and separation properties is likely to take place in the near future. Thanks to the enormous success of MOFs, gas sorption analysis has become a mainstream technique that can be readily employed for systematically assessing the potential of microporous MPs for a range of gas separations, especially those where challenging conditions can limit the applicability of carboxylate-based MOFs.

4.3. Electrochemical Devices

While global energy demand grows year on year, with a projected 30% increase by the year 2040 [120] there has been a lot of interest in finding new, more efficient, and cleaner ways of powering our future. In this context, electrochemical devices such as rechargeable batteries and fuel cells are becoming increasingly important for energy storage and conversion, attracting considerable research efforts [121,122]. Technically, a battery and a fuel cell are both electrochemical cells, consisting of an anodic and a cathodic compartment, which are connected through an electrical circuit that allows the exchange of electrons, while an electrolyte ensures the mobility of positive charge carriers (Figure 29) [123]. The main difference between the two devices is that rechargeable batteries can reversibly store electricity in the form of chemical energy and release it on demand, whereas fuel cells convert the chemical energy contained in a fuel (typically hydrogen) into electricity. The interest in maximising the performance of these devices is currently a major drive in developing new materials that can serve as either electrodes or electrolyte components [123].

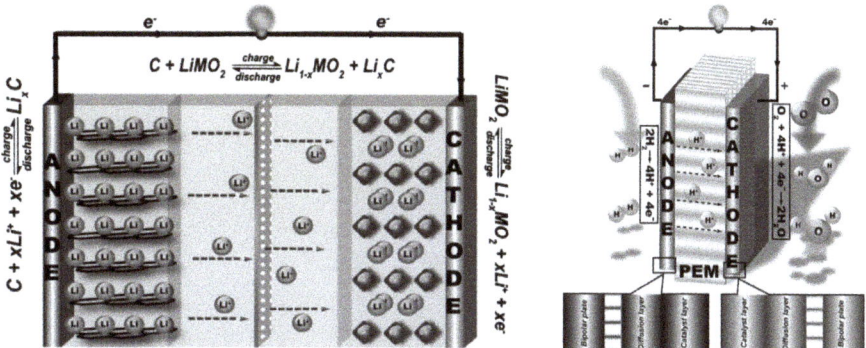

Figure 29. Schematic representations of a rechargeable lithium ion (Li-ion) battery (**left**) and a proton exchange membrane fuel cell (**right**). Adapted with permission from ref [123]. Copyright 2009, Elsevier Inc.

4.3.1. Solid-State Proton Conductors for Fuel Cells

The positive charge carriers in fuel cells are protons, which need to be efficiently transported via the electrolyte from the anode, where H_2 is oxidised, to the cathode, where they combine with oxygen to form water, in order to maximise the power generation. This makes the electrolyte a crucial component in determining the efficiency of a fuel cell, driving research into materials displaying high proton conductivity [124,125]. The most common type of fuel cell is the proton exchange membrane fuel cells (PEMFC), where the electrolyte is typically a proton conductive polymer, such as Nafion, polybenzimidazole, and sulfonated polyether-ether ketones. In spite of their very high conductivity, these polymers often suffer from limited thermal and mechanical stability, which affects their long-term performance. As early as the 1990s, layered zirconium phosphonates were proposed as alternative proton conduction materials [126,127], owing to the ability to functionalise them with groups of variable acid strength. In particular, zirconium sulfophenylphosphonate was found to have excellent proton conductivity in humid conditions, reaching values as high as 1.6×10^{-2} S·cm^{-1}. The application of MPs as proton conductors has recently been extensively reviewed [6,128]; therefore, we will only present some selected examples.

A recent example of a metal phosphonate with good proton conduction is that of La(H$_5$DTMP)· 7H$_2$O, which is based on the hexamethylenediamine-N,N,N',N'-tetrakis(methylenephosphonic acid) linker (Figure 30a) [129]. Structural analysis of the compound showed that it is a three-dimensional (3D) framework featuring narrow one-dimensional (1D) channels where seven water molecules per formula unit are accommodated, forming a network of hydrogen bonds involving non-coordinated P–O groups that extend throughout the channels (Figure 30b). This allows for efficient proton conduction to take place, as proved by the conductivity value of 8×10^{-3} S cm^{-1} at 25 °C and 98% relative humidity, as measured by impedance analysis. The activation energy was found to be 0.23 eV, which is typical of a Grotthuss-type mechanism, where well-ordered water molecules play a key role in enabling efficient proton shuttling. Other MPs with similar "water channels" were subsequently reported [130–133], which displayed good proton conduction and activation energies consistent with the Grotthuss mechanism.

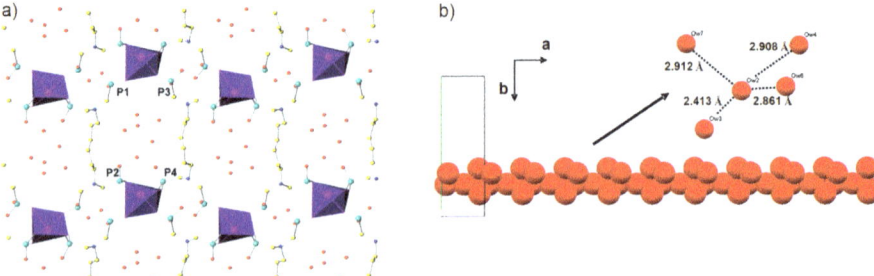

Figure 30. (a) Crystal structure of La(H$_5$DTMP)·7H$_2$O and (b) the extended network of hydrogen-bonded water molecules along the *a*-axis, along with a detail of the hydrogen bonding distances. Colour code: lanthanum: purple, phosphorus: light blue, oxygen: red, nitrogen: dark blue, carbon: yellow. Adapted with permission from reference [129]. Copyright 2012, Royal Society of Chemistry.

These results highlight the importance of having well-defined pathways for the protons to hop between charge carriers to maximise conductivity. Clearly, knowledge of the crystal structure is essential to be able to identify such pathways in the first place. It is also worth emphasising that using polyphosphonic acids as linkers is crucial, since this increases the chances of having free P–OH groups that can facilitate the formation of extended hydrogen bond networks with water molecules. These characteristics can be enhanced by post-synthesis modifications, for instance by the adsorption of molecules, which contribute to generate robust proton transfer pathways [130].

4.3.2. Electrodes for Rechargeable Batteries

Different from fuel cells, the efficiency of a rechargeable battery not only depends on the mobility of the charge carriers, but heavily relies on the ability of the electrodes to reversibly store and release these charge carriers during charging and discharging. Typical charge carriers are monovalent alkaline ions, with lithium element presently being the most employed, owing to its small size and low weight, which ensure high charge density. In order to maximise the performance, it is important that both electrodes possess high specific capacity and long-term stability. The state of the art of the electrodes for lithium ion (Li-ion) batteries includes graphite as the anodic material (which is present in the negative electrode, theoretical specific capacity = 372 mAh·g^{-1}) and LiCoO$_2$ as the cathodic material (which is present in the positive electrode, theoretical specific capacity = 137 mAh·g^{-1}). They are both layered compounds where Li-ions are inserted and extracted by intercalation and deintercalation processes, respectively. The major limitation of the energy density of present Li-ion batteries is the relatively low specific capacity of the cathode. In addition, the latter is slowly decreasing during cycling due to the material instability (Co dissolution in the electrolyte) and other mechanisms.

Recently, a MP based on Fe(II) and methylenediphosphonic acid, of formula Li$_{1.4}$Fe$_{6.8}$[CH$_2$(PO$_3$)$_2$]$_3$[CH$_2$(PO$_3$)(PO$_3$H)]·4H$_2$O, was extensively studied for its potential as a positive electrode for Li-ion batteries (Figure 31a,b) [134]. Structural characterisation was carried out combining synchrotron PXRD and neutron diffraction data, finding that the compound has a very similar structure to a previously reported Co methylenediphosphonate [30]. Three different crystallographic sites were identified for Fe(II) ions: two of them are octahedrally coordinated, while the other one is tetrahedrally coordinated. The compound was prepared in the presence of Li, which was retained within the crystal structure as an extra-framework species that was accommodated in the small 1D channels, thus rendering a pre-lithiated electrode. Electrochemical testing showed a specific charge of 85 mAh·g^{-1} until the 60th cycle, corresponding to 50% of the theoretical value of 168 mAh·g^{-1}. The origin of this lower value was attributed to part of the Fe ions not undergoing any redox process, as evidenced by in situ near-edge X-ray absorption spectroscopy. Ex situ XRD analysis after the first electrochemical cycle showed that there was no significant loss of crystallinity, suggesting that activation did not proceed via decomposition of the compound. Successive studies focused on testing the same type of framework as a negative electrode, using both Fe and Co as the metal component [135], and combining methylenediphosphonic acid with Ni, which leads to the formation of different crystal structures [136]. In all cases, it was found that the structure irreversibly amorphises after the first cycle due to a conversion reaction mechanism involving the extrusion of the transition metal(II) as nanoparticles.

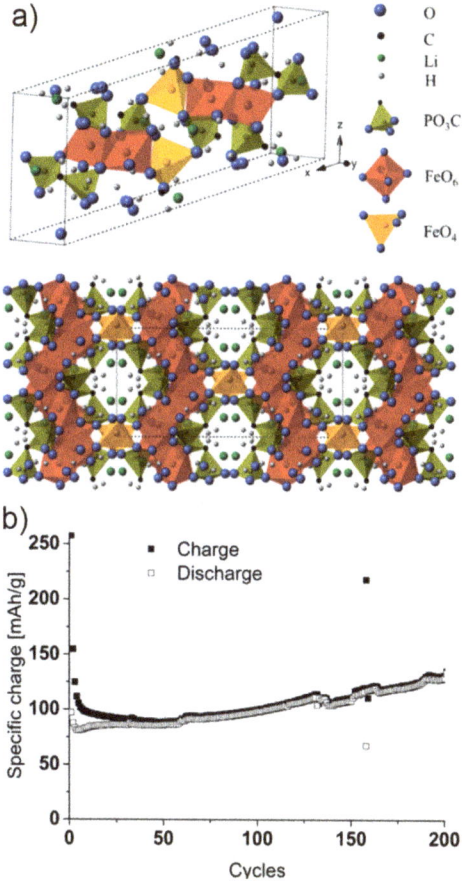

Figure 31. (a) Illustration of the monoclinic unit cell of $Li_{1.4}Fe_{6.8}[CH_2(PO_3)_2]_3[CH_2(PO_3)(PO_3H)]\cdot 4H_2O$ and the expanded crystal structure viewed along the c-axis. Colour code: octahedral iron: red, tetrahedral iron: orange, phosphorus: yellow, oxygen: blue, carbon: black, water: grey, and lithium: green. (b) Specific charge vs. cycle number plot of $Li_{1.4}Fe_{6.8}[CH_2(PO_3)_2]_3[CH_2(PO_3)(PO_3H)]\cdot 4H_2O$ cycled at 20 mA/g with a 1-h potentiostatic step after each half-cycle. Adapted with permission from reference [134]. Copyright 2015, American Chemical Society.

The compounds discussed above represent some of the first reported examples of MPs being applied as electrode materials for rechargeable batteries. There is a growing interest in employing hybrid organic–inorganic materials for this type of application, mainly because of the possibility to tailor their structure, something that is not easily done when dealing with purely inorganic materials. Given the structural versatility of MPs and their high stability, there is great potential in this growing area of research.

4.4. Drug Delivery

Phosphonic acids (most often in their sodium phosphonate form) have featured in the treatment of bone-related conditions, e.g., osteoporosis and Paget's disease, for more than 30 years. Specifically, bisphosphonates (BPs) have been used as stable analogues of pyrophosphate, in which the P–O–P bonds are replaced with P–C–P bonds, making them more stable and less prone to enzymatic attack [137]. The way in which BPs can help in the treatment of bone-related conditions is through selective adsorption to the bone surface, leading to the interruption of bone resorption, whereby osteoclasts are prevented from resorbing bone tissue. There are two classes of BPs that can be used here: non-N-containing compounds, such as etidronate and clodronate, and N-containing compounds, such as pamidronate and alendronate. Perhaps one of the main drawbacks for the use of BPs for treating conditions such as osteoporosis is the limited bioavailability when delivered orally [138]. This means that in order to obtain the desired therapeutic effect, it is often the case that the drug dosage is increased. While this appears to overcome the bioavailability issue, it can also lead to an increase in other known side effects, i.e., hypocalcaemia, oesophageal cancer, and atrial fibrillation, etc. In order to overcome this issue, delayed release (DR) administration and other methods have been developed, helping to negate the need to increase the dosage. An example of one of the DR drugs is risedronate 35 mg (DR). Unfortunately, these drugs still suffer in that they require long periods of fasting due to the complexing that can occur with food in the stomach.

In a paper published in 2019, Papathanasiou et al. explored various routes for the controlled release of bisphosphonates, one of which was based on self-sacrificial MOFs [139]. Using etidronic acid (ETID, one of the "early" BPs) as the linker, the researchers synthesised Ca–ETID and presented it in tablet form, comparing it to a control containing "free" ETID. The structure of Ca–ETID exists as a 1D chain, where each Ca^{2+} centre is coordinated to three ETID linkers, the –OH group, and water (Figure 32). Then, these single chains form 'double chains' held together by a network of hydrogen bonds between ETID linkers. The release rates of the control and Ca–ETID tablets upon soaking in an aqueous solution at pH 1.3 (representative of the environment in the stomach) were determined by NMR spectroscopy, finding that the control showed 100% release after just 10 h, whereas Ca-ETID showed only 30% release in the same time period, reaching a maximum of 90% after 150 h (Figure 32). The slower release rate of Ca–ETID tablets was ascribed to the slow rate of hydrolysis of the metal–phosphonate coordination bonds in the solvent medium.

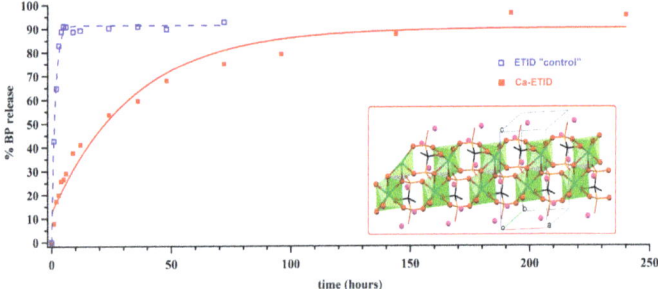

Figure 32. Etidronic acid (ETID) release versus time for the control tablet (dark blue circles) and the Ca–ETID tablet (light blue triangles). The crystal structure of Ca–ETID is also displayed in the inset. Colour code: calcium: green, phosphorus: yellow, oxygen: red, carbon: black, lattice water: pink. Adapted with permission from reference [139]. Copyright 2018, Walter de Gruyter GmbH and Co. KG.

These promising results are part of ongoing work that aims at systematically investigating the preparation and structural characterisation of a series of MPs based on various BP and biocompatible metals (e.g., Mg, Ca, Sr, and Ba), as well as their performance as matrices for controlled drug delivery. The final goal is to identify clear structure/activity relationships between specific structural features, such as the metal–phosphonate bond strength, the metal ion radius, the existence of supramolecular interactions, and the drug delivery efficacy.

5. Outlook

Forty years after its inception, metal phosphonate chemistry has come a long way, but the breadth of topics discussed during the First European Workshop on Metal Phosphonate Chemistry, and reviewed here, shows that the field is still very lively and moving in new directions. From the synthesis standpoint, we have highlighted the potential of high-throughput methods in accelerating the discovery of new materials, the ability of mechanochemistry to provide access to structures not obtainable through conventional synthesis, and the intense efforts in developing permanently porous materials, taking inspiration from MOF chemistry, including the first demonstration of the suitability of phosphinic acids to serve as alternative building blocks for highly stable MOFs. On the characterisation side, the main advances discussed relate to the development of methods for structural solutions of microcrystalline and nanocrystalline compounds using electron diffraction, which might help overcome what is probably the longest standing challenge in the field. Furthermore, the employment of in situ methods can identify the different stages of the crystallisation process, which is knowledge that can be crucial to designing better synthetic processes. Looking at applications, the stability and structural versatility of MPs makes them suitable for employment in several fields: in catalysis, they can either serve as effective supports for catalytic nanoparticles or display intrinsic catalytic properties; the development of microporous phosphonate-based frameworks is also promoting investigation towards their employment in gas separations, as an alternative to other MOF classes; the ever-growing interest in developing new electrochemical devices is driving research aimed at evaluating their potential as both solid-state proton conductors and redox active materials; finally, the complexation of bisphosphonic drugs with alkaline earth metals shows promise as an effective strategy for controlled drug delivery. Our hope is that the First European Workshop on Metal Phosphonates Chemistry will serve as a sort of nucleation point for promoting the growth of a larger network of collaborations within the field, both at the European level and beyond, which would be crucial in order to make the next 40 years of metal phosphonates chemistry as productive and successful as the past 40 have been. A second edition of the workshop is taking place in Berlin in September 2019, demonstrating that there is indeed a strong interest within the community to keep the conversation open and look ahead to the future.

Author Contributions: Conceptualization, S.J.I.S. and M.T.; writing—original draft preparation, S.J.I.S. and M.T.; writing—review and editing, all authors.

Funding: M.T is supported by funding from the European Union's Horizon 2020 research and innovation program under the Marie Skłodowska-Curie grant agreement No 663830.

Conflicts of Interest: The authors declare no conflict of interest.

References

1. Clearfield, A.; Demadis, K. (Eds.) *Metal Phosphonate Chemistry*; Royal Society of Chemistry: Cambridge, UK, 2011; ISBN 978-1-84973-356-4.
2. Clearfield, A. Layered Phosphates, Phosphites and Phosphonates of Groups 4 and 14 Metals. *Comments Inorg. Chem.* **1990**, *10*, 89–128. [CrossRef]
3. Cao, G.; Mallouk, T.E. Shape-Selective Intercalation Reactions of Layered Zinc and Cobalt Phosphonates. *Inorg. Chem.* **1991**, *30*, 1434–1438. [CrossRef]
4. Zhang, Y.; Clearfield, A. Synthesis, Crystal Structures, and Coordination Intercalation Behavior of Two Copper Phosphonates. *Inorg. Chem.* **1992**, *25*, 2821–2826. [CrossRef]
5. Zhang, B.; Poojary, D.M.; Clearfield, A.; Peng, G. Synthesis, Characterization, and Amine Intercalation Behavior of Zirconium N-(Phosphonomethyl)iminodiacetic Acid Layered Compounds. *Chem. Mater.* **2002**, *8*, 1333–1340. [CrossRef]
6. Bao, S.S.; Shimizu, G.K.H.; Zheng, L.M. Proton conductive metal phosphonate frameworks. *Coord. Chem. Rev.* **2019**, *378*, 577–594. [CrossRef]

7. Curini, M.; Rosati, O.; Costantino, U. Heterogeneous Catalysis in Liquid Phase Organic Synthesis, Promoted by Layered Zirconium Phosphates and Phosphonates. *Curr. Org. Chem.* **2005**, *8*, 591–606. [CrossRef]
8. Zhu, Y.P.; Ma, T.Y.; Liu, Y.L.; Ren, T.Z.; Yuan, Z.Y. Metal phosphonate hybrid materials: From densely layered to hierarchically nanoporous structures. *Inorg. Chem. Front.* **2014**, *1*, 360–383. [CrossRef]
9. Goura, J.; Chandrasekhar, V. Molecular Metal Phosphonates. *Chem. Rev.* **2015**, *115*, 6854–6965. [CrossRef]
10. Vivani, R.; Alberti, G.; Costantino, F.; Nocchetti, M. New advances in zirconium phosphate and phosphonate chemistry: Structural archetypes. *Microporous Mesoporous Mater.* **2008**, *107*, 58–70. [CrossRef]
11. Bao, S.S.; Zheng, L.M. Magnetic materials based on 3D metal phosphonates. *Coord. Chem. Rev.* **2016**, *319*, 63–85. [CrossRef]
12. Shimizu, G.K.H.; Vaidhyanathan, R.; Taylor, J.M. Phosphonate and sulfonate metal organic frameworks. *Chem. Soc. Rev.* **2009**, *38*, 1430–1449. [CrossRef]
13. Gagnon, K.J.; Perry, H.P.; Clearfield, A. Conventional and unconventional metal-organic frameworks based on phosphonate ligands: MOFs and UMOFs. *Chem. Rev.* **2012**, *112*, 1034–1054. [CrossRef]
14. Yücesan, G.; Zorlu, Y.; Stricker, M.; Beckmann, J. Metal-organic solids derived from arylphosphonic acids. *Coord. Chem. Rev.* **2018**, *369*, 105–122. [CrossRef]
15. Clearfield, A.; David Smith, G. The Crystallography and Structure of α-zirconium bis(monohydrogen orthophosphate) monohydrate. *Inorg. Chem.* **1969**, *8*, 431–436. [CrossRef]
16. Alberti, G.; Costantino, U.; Allulli, S.; Tomassini, N. Crystalline $Zr(R-PO_3)_2$ and $Zr(R-OPO_3)_2$ compounds (R = organic radical). A new class of materials having layered structure of the zirconium phosphate type. *J. Inorg. Nucl. Chem.* **1978**, *40*, 1113–1117. [CrossRef]
17. Alberti, G.; Casciola, M.; Costantino, U.; Vivani, R. Layered and pillared metal(IV) phosphates and phosphonates. *Adv. Mater.* **1996**, *8*, 291–303. [CrossRef]
18. Vivani, R.; Costantino, F.; Taddei, M. Zirconium Phosphonates. In *Metal Phosphonate Chemistry*; Clearfield, A., Demadis, K.D., Eds.; Royal Society of Chemistry: Cambridge, UK, 2011; pp. 45–86.
19. Cunningham, D.; Hennelly, P.J.D.; Deeney, T. Divalent metal phenylphosphonates and phenylarsonates. *Inorg. Chim. Acta* **1979**, *37*, 95–102. [CrossRef]
20. Cao, G.; Lee, H.; Lynch, V.M.; Mallouk, T.E. Structural Studies of some New Lamellar Magnesium, Manganese and Calcium Phosphonates. *Solid State Ion.* **1988**, *26*, 63–69. [CrossRef]
21. Cao, G.; Lee, H.; Lynch, V.M.; Mallouk, T.E. Synthesis and Structural Characterization of a Homologous Series of Divalent-Metal Phosphonates, $M^{II}(O_3PR) \cdot H_2O$ and $M^{II}(HO_3PR)_2$. *Inorg. Chem.* **1988**, *27*, 2781–2785. [CrossRef]
22. Martin, K.J.; Squattrito, P.J.; Clearfield, A. The crystal and molecular structure of Zinc Phenylphosphonate. *Carbohydr. Res.* **1989**, *194*, 7–9. [CrossRef]
23. Poojary, M.D.; Hu, H.L.; Campbell, F.L.; Clearfield, A. Determination of crystal structures from limited powder data sets: Crystal structure of zirconium phenylphosphonate. *Acta Crystallogr. Sect. B* **1993**, *49*, 996–1001. [CrossRef]
24. Maeda, K.; Kiyozumi, Y.; Mizukami, F. Synthesis of the First Microporous Aluminum Phosphonate with Organic Groups Covalently Bonded to the Skeleton. *Angew. Chem. Int. Ed.* **1994**, *33*, 2335–2337. [CrossRef]
25. Le Bideau, J.; Payen, C.; Palvadeau, P.; Bujoli, B. Preparation, Structure, and Magnetic Properties of Copper(II) Phosphonates. β-$Cu^{II}(CH_3PO_3)$, an Original Three-Dimensional Structure with a Channel-Type Arrangement. *Inorg. Chem.* **1994**, *33*, 4885–4890. [CrossRef]
26. Maeda, K.; Akimoto, J.; Kiyozumi, Y.; Mizukami, F. AlMepO-α: A Novel Open–Framework Aluminum Methylphosphonate with Organo-Lined Unidimensional Channels. *Angew. Chem. Int. Ed.* **1995**, *34*, 1199–1201. [CrossRef]
27. Drumel, S.; Janvier, P.; Deniaud, D.; Bujoli, B. Synthesis and crystal structure of $Zn(O_3PC_2H_4NH_2)$, the first functionalized zeolite-like phosphonate. *J. Chem. Soc. Chem. Commun.* **1995**, 1051–1052. [CrossRef]
28. Maeda, K.; Akimoto, J.; Kiyozumi, Y.; Mizukami, F. Structure of aluminium methylphosphonate, AlMepO-β, with unidimensional channels formed from ladder-like organic-inorganic polymer chains. *J. Chem. Soc. Chem. Commun.* **1995**, 1033–1034. [CrossRef]
29. Maeda, K. Metal phosphonate open-framework materials. *Microporous Mesoporous Mater.* **2004**, *73*, 47–55. [CrossRef]

30. Lohse, D.L.; Sevov, S.C. Co$_2$(O$_3$P-CH$_2$-PO$_3$)·H$_2$O: A Novel Microporous Diphosphonate with an Inorganic Framework and Hydrocarbon-Lined Hydrophobic Channels. *Angew. Chem. Int. Ed.* **1997**, *36*, 1619–1621. [CrossRef]
31. Hix, G.B.; Kariuki, B.M.; Kitchin, S.; Tremayne, M. Synthesis and structural characterization of Zn(O$_3$PCH$_2$OH), a new microporous zinc phosphonate. *Inorg. Chem.* **2001**, *40*, 1477–1481. [CrossRef]
32. Mason, M.R. Molecular Phosphates, Phosphonates, Phosphinates, and Arsonates of the Group 13 Elements. *J. Clust. Sci.* **1998**, *9*, 1–23. [CrossRef]
33. Walawalkar, M.G.; Roesky, H.W.; Murugavel, R. Molecular phosphonate cages: Model compounds and starting materials for phosphate materials. *Acc. Chem. Res.* **1999**, *32*, 117–126. [CrossRef]
34. Gopal, K.; Ali, S.; Winpenny, R.E.P. Structural Studies of Paramagnetic Molecular Phosphonates. In *Metal Phosphonate Chemistry*; Royal Society of Chemistry: Cambridge, UK, 2011; pp. 364–419.
35. Chandrasekhar, V.; Senapati, T.; Dey, A.; Hossain, S. Molecular transition-metal phosphonates. *Dalton Trans.* **2011**, *40*, 5394–5418. [CrossRef] [PubMed]
36. Mason, M.R.; Mashuta, M.S.; Richardson, J.F. Cyclic and Cubic Organophosphonates of Gallium and Their Relationship to Structural Motifs in Gallophosphate Molecular Sieves. *Angew. Chem. Int. Ed.* **1997**, *36*, 239–241. [CrossRef]
37. Clearfield, A. The Early History and Growth of Metal Phosphonate Chemistry. In *Metal Phosphonate Chemistry*; Royal Society of Chemistry: Cambridge, UK, 2011; pp. 1–44.
38. Stock, N. High-throughput investigations employing solvothermal syntheses. *Microporous Mesoporous Mater.* **2010**, *129*, 287–295. [CrossRef]
39. Bauer, S.; Stock, N. Implementation of a temperature-gradient reactor system for high-throughput investigation of phosphonate-based inorganic-organic hybrid compounds. *Angew. Chem. Int. Ed.* **2007**, *46*, 6857–6860. [CrossRef] [PubMed]
40. Stock, N.; Bein, T. High-Throughput Synthesis of Phosphonate-Based Inorganic-Organic Hybrid Compounds under Hydrothermal Conditions. *Angew. Chem. Int. Ed.* **2004**, *43*, 749–752. [CrossRef]
41. Bauer, S.; Bein, T.; Stock, N. High-throughput investigation and characterization of cobalt carboxy phosphonates. *Inorg. Chem.* **2005**, *44*, 5882–5889. [CrossRef]
42. Forster, P.M.; Stock, N.; Cheetham, A.K. A high-throughput investigation of the role of pH, temperature, concentration, and time on the synthesis of hybrid inorganic-organic materials. *Angew. Chem. Int. Ed.* **2005**, *44*, 7608–7611. [CrossRef]
43. Stock, N.; Bein, T. High-throughput investigation of metal carboxyarylphosphonate hybrid compounds. *J. Mater. Chem.* **2005**, *15*, 1384–1391. [CrossRef]
44. Maniam, P.; Stock, N. High-throughput Methods for the Systematic Investigation of Metal Phosphonate Synthesis Fields. In *Metal Phosphonate Chemistry*; Royal Society of Chemistry: Cambridge, UK, 2011; pp. 87–106.
45. Hermer, N.; Stock, N. The new triazine-based porous copper phosphonate [Cu$_3$(PPT)(H$_2$O)$_3$]·10H$_2$O. *Dalton Trans.* **2015**, *44*, 3720–3723. [CrossRef]
46. Rhauderwiek, T.; Wolkersdörfer, K.; Øien-Ødegaard, S.; Lillerud, K.P.; Wark, M.; Stock, N. Crystalline and permanently porous porphyrin-based metal tetraphosphonates. *Chem. Commun.* **2018**, *54*, 389–392. [CrossRef] [PubMed]
47. Rhauderwiek, T.; Stock, N.; Bueken, B.; Döblinger, M.; Wuttke, S.; Zhao, H.; Reinsch, H.; Hirschle, P.; De Vos, D.; Kolb, U. Highly stable and porous porphyrin-based zirconium and hafnium phosphonates—Electron crystallography as an important tool for structure elucidation. *Chem. Sci.* **2018**, *9*, 5467–5478. [CrossRef]
48. Stock, N.; Rauscher, M.; Bein, T. Inorganic-organic hybrid compounds: Hydrothermal synthesis and characterization of a new three-dimensional metal tetraphosphonate Mn[(HO$_3$PCH$_2$)N(H)(CH$_2$)$_4$(H)N(CH$_2$PO$_3$H)$_2$]. *J. Solid State Chem.* **2004**, *177*, 642–647. [CrossRef]
49. Hermer, N.; Reinsch, H.; Mayer, P.; Stock, N. Synthesis and characterisation of the porous zinc phosphonate [Zn$_2$(H$_2$PPB)(H$_2$O)$_2$]·xH$_2$O. *CrystEngComm* **2016**, *18*, 8147–8150. [CrossRef]
50. Feyand, M.; Seidler, C.F.; Deiter, C.; Rothkirch, A.; Lieb, A.; Wark, M.; Stock, N. High-throughput microwave-assisted discovery of new metal phosphonates. *Dalton Trans.* **2013**, *42*, 8761–8770. [CrossRef]
51. Schilling, L.H.; Stock, N. High-throughput ultrasonic synthesis and in situ crystallisation investigation of metal phosphonocarboxylates. *Dalton Trans.* **2014**, *43*, 414–422. [CrossRef] [PubMed]

52. Boldyrev, V.; Tkacova, K. Mechanochemistry of Solids: Past, Present, and Prospects. *J. Mater. Synth. Process.* **2000**, *8*, 121–122. [CrossRef]
53. Takacs, L. The historical development of mechanochemistry. *Chem. Soc. Rev.* **2013**, *42*, 7649–7659. [CrossRef] [PubMed]
54. Bruckmann, A.; Krebs, A.; Bolm, C. Organocatalytic reactions: Effects of ball milling, microwave and ultrasound irradiation. *Green Chem.* **2008**, *10*, 1131–1141. [CrossRef]
55. Howard, J.L.; Cao, Q.; Browne, D.L. Mechanochemistry as an emerging tool for molecular synthesis: What can it offer? *Chem. Sci.* **2018**, *9*, 3080–3094. [CrossRef]
56. Wilke, M.; Batzdorf, L.; Fischer, F.; Rademann, K.; Emmerling, F. Cadmium phenylphosphonates: Preparation, characterisation and in situ investigation. *RSC Adv.* **2016**, *6*, 36011–36019. [CrossRef]
57. Wilke, M.; Buzanich, A.G.; Reinholz, U.; Rademann, K.; Emmerling, F. The structure and In Situ synthesis investigation of isomorphic mononuclear molecular metal phenylphosphonates. *Dalton Trans.* **2016**, *45*, 9460–9467. [CrossRef]
58. Wilke, M.; Kabelitz, A.; Zimathies, A.; Rademann, K.; Emmerling, F. Crystal structure and in situ investigation of a mechanochemical synthesized 3D zinc N-(phosphonomethyl)glycinate. *J. Mater. Sci.* **2017**, *52*, 12013–12020. [CrossRef]
59. Akhmetova, I.; Schutjajew, K.; Wilke, M.; Buzanich, A.; Rademann, K.; Roth, C.; Emmerling, F. Synthesis, characterization and in situ monitoring of the mechanochemical reaction process of two manganese(II)-phosphonates with N-containing ligands. *J. Mater. Sci.* **2018**, *53*, 13390–13399. [CrossRef]
60. Maeda, K.; Kiyozumi, Y.; Mizukami, F. Characterization and Gas Adsorption Properties of Aluminum Methylphosphonates with Organically Lined Unidimensional Channels. *J. Phys. Chem. B* **2002**, *101*, 4402–4412. [CrossRef]
61. Li, H.; Eddaoudi, M.; O'Keeffe, M.; Yaghi, O.M. Design and synthesis of an exceptionally stable and highly porous metal-organic framework. *Nature* **1999**, *402*, 276–279. [CrossRef]
62. Chui, S.S.Y.; Lo, S.M.F.; Charmant, J.P.H.; Orpen, A.G.; Williams, I.D. A chemically functionalizable nanoporous material $[Cu_3(TMA)_2(H_2O)_3]_{(n)}$. *Science* **1999**, *283*, 1148–1150. [CrossRef]
63. Marsolier, G.; Serre, C.; Thouvenot, C.; Noguès, M.; Louër, D.; Millange, F.; Férey, G. Very Large Breathing Effect in the First Nanoporous Chromium(III)-Based Solids: MIL-53 or $Cr^{III}(OH)\cdot\{O_2C-C_6H_4-CO_2\}\cdot\{HO_2C-C_6H_4-CO_2H\}_x\cdot H_2O_y$. *J. Am. Chem. Soc.* **2002**, *124*, 13519–13526. [CrossRef]
64. Long, J.R.; Yaghi, O.M. The pervasive chemistry of metal–organic frameworks. *Chem. Soc. Rev.* **2009**, *38*, 1213. [CrossRef] [PubMed]
65. Zhou, H.-C.; Long, J.R.; Yaghi, O.M. Introduction to Metal–Organic Frameworks. *Chem. Rev.* **2012**, *112*, 673–674. [CrossRef] [PubMed]
66. Rueff, J.M.; Hix, G.B.; Jaffrès, P.A. Rigid Phosphonic Acids as Building Blocks for Crystalline Hybrid Materials. In *Tailored Organic-Inorganic Materials*; John Wiley & Sons, Inc: Hoboken, NJ, USA, 2015; pp. 341–393, ISBN 9781118792223.
67. Serre, C.; Groves, J.A.; Lightfoot, P.; Slawin, A.M.Z.; Wright, P.A.; Stock, N.; Bein, T.; Haouas, M.; Taulelle, F.; Férey, G. Synthesis, structure and properties of related microporous N,N'-piperazinebismethylenephosphonates of aluminum and titanium. *Chem. Mater.* **2006**, *18*, 1451–1457. [CrossRef]
68. Benoit, V.; Pillai, R.S.; Orsi, A.; Normand, P.; Jobic, H.; Nouar, F.; Billemont, P.; Bloch, E.; Bourrelly, S.; Devic, T.; et al. MIL-91(Ti), a small pore metal-organic framework which fulfils several criteria: An upscaled green synthesis, excellent water stability, high CO_2 selectivity and fast CO_2 transport. *J. Mater. Chem. A* **2016**, *4*, 1383–1389. [CrossRef]
69. Llewellyn, P.L.; Garcia-Rates, M.; Gaberová, L.; Miller, S.R.; Devic, T.; Lavalley, J.C.; Bourrelly, S.; Bloch, E.; Filinchuk, Y.; Wright, P.A.; et al. Structural origin of unusual CO_2 adsorption behavior of a small-pore aluminum bisphosphonate MOF. *J. Phys. Chem. C* **2015**, *119*, 4208–4216. [CrossRef]
70. Groves, J.A.; Miller, S.R.; Warrender, S.J.; Mellot-Draznieks, C.; Lightfoot, P.; Wright, P.A. The first route to large pore metal phosphonates. *Chem. Commun.* **2006**, *1*, 3305–3307. [CrossRef] [PubMed]

71. Wharmby, M.T.; Pearce, G.M.; Mowat, J.P.S.; Griffin, J.M.; Ashbrook, S.E.; Wright, P.A.; Schilling, L.-H.; Lieb, A.; Stock, N.; Chavan, S.; et al. Synthesis and crystal chemistry of the STA-12 family of metal N,N′-piperazinebis(methylenephosphonate)s and applications of STA-12(Ni) in the separation of gases. *Microporous Mesoporous Mater.* **2011**, *157*, 3–17. [CrossRef]
72. Wharmby, M.T.; Mowat, J.P.S.; Thompson, S.P.; Wright, P.A. Extending the pore size of crystalline metal phosphonates toward the mesoporous regime by isoreticular synthesis. *J. Am. Chem. Soc.* **2011**, *133*, 1266–1269. [CrossRef]
73. Taddei, M.; Costantino, F.; Vivani, R. Robust Metal-Organic Frameworks Based on Tritopic Phosphonoaromatic Ligands. *Eur. J. Inorg. Chem.* **2016**, *2016*, 4300–4309. [CrossRef]
74. Firmino, A.D.G.; Figueira, F.; Tomé, J.P.C.; Paz, F.A.A.; Rocha, J. Metal–Organic Frameworks assembled from tetraphosphonic ligands and lanthanides. *Coord. Chem. Rev.* **2018**, *355*, 133–149. [CrossRef]
75. Zaręba, J.K. Tetraphenylmethane and tetraphenylsilane as building units of coordination polymers and supramolecular networks—A focus on tetraphosphonates. *Inorg. Chem. Commun.* **2017**, *86*, 172–186. [CrossRef]
76. Zheng, T.; Yang, Z.; Gui, D.; Liu, Z.; Wang, X.; Dai, X.; Liu, S.; Zhang, L.; Gao, Y.; Chen, L.; et al. Overcoming the crystallization and designability issues in the ultrastable zirconium phosphonate framework system. *Nat. Commun.* **2017**, *8*, 15369. [CrossRef]
77. Erkal, T.S.; Bulut, A.; Zorlu, Y.; Beckmann, J.; Yücesan, G.; Erbahar, D.; Yazaydin, A.O.; Çetinkaya, A. A cobalt arylphosphonate MOF—Superior stability, sorption and magnetism. *Chem. Commun.* **2019**, *55*, 3053–3056. [CrossRef]
78. Wang, B.; Huang, Z.; Yang, T.; Rhauderwiek, T.; Xu, H.; Stock, N.; Inge, A.K.; Zou, X. A Porous Cobalt Tetraphosphonate Metal-Organic Framework: Accurate Structure and Guest Molecule Location Determined by Continuous-Rotation Electron Diffraction. *Chem. Eur. J.* **2018**, *24*, 17429–17433. [CrossRef] [PubMed]
79. Carson, I.; Healy, M.R.; Doidge, E.D.; Love, J.B.; Morrison, C.A.; Tasker, P.A. Metal-binding motifs of alkyl and aryl phosphinates; versatile mono and polynucleating ligands. *Coord. Chem. Rev.* **2017**, *335*, 150–171. [CrossRef]
80. Costantino, F.; Ienco, A.; Taddei, M. Hybrid Multifunctional Materials Based on Phosphonates, Phosphinates and Auxiliary Ligands. In *Tailored Organic-Inorganic Materials*; John Wiley & Sons, Inc.: Hoboken, NJ, USA, 2015; pp. 193–244, ISBN 9781118792223.
81. Shimizu, G.K.H.; Martens, I.; Woo, T.K.; Vaidhyanathan, R.; Daff, T.D.; Yeganegi, S.; Aghaji, M.Z.; Iremonger, S.S.; Liang, J. Phosphonate Monoesters as Carboxylate-like Linkers for Metal Organic Frameworks. *J. Am. Chem. Soc.* **2011**, *133*, 20048–20051. [CrossRef]
82. Hynek, J.; Brázda, P.; Rohlíček, J.; Londesborough, M.G.S.; Demel, J. Phosphinic Acid Based Linkers: Building Blocks in Metal–Organic Framework Chemistry. *Angew. Chem. Int. Ed.* **2018**, *57*, 5016–5019. [CrossRef]
83. Taylor, J.M.; Vaidhyanathan, R.; Iremonger, S.S.; Shimizu, G.K.H. Enhancing water stability of metal-organic frameworks via phosphonate monoester linkers. *J. Am. Chem. Soc.* **2012**, *134*, 14338–14340. [CrossRef] [PubMed]
84. Gelfand, B.S.; Lin, J.B.; Shimizu, G.K.H. Design of a Humidity-stable metal-organic framework using a phosphonate monoester ligand. *Inorg. Chem.* **2015**, *54*, 1185–1187. [CrossRef]
85. Taddei, M.; Costantino, F.; Vivani, R. Synthesis and crystal structure from X-ray powder diffraction data of Two zirconium diphosphonates containing piperazine groups. *Inorg. Chem.* **2010**, *49*, 9664–9670. [CrossRef]
86. Gonen, T.; Hattne, J.; Rodriguez, J.A.; Martynowycz, M.W.; Stoltz, B.M.; Fulton, T.J.; Jones, C.G.; Nelson, H.M. The CryoEM Method MicroED as a Powerful Tool for Small Molecule Structure Determination. *ACS Cent. Sci.* **2018**, *4*, 1587–1592. [CrossRef]
87. Gruene, T.; Wennmacher, J.T.C.; Zaubitzer, C.; Holstein, J.J.; Heidler, J.; Fecteau-Lefebvre, A.; De Carlo, S.; Müller, E.; Goldie, K.N.; Regeni, I.; et al. Rapid Structure Determination of Microcrystalline Molecular Compounds Using Electron Diffraction. *Angew. Chem. Int. Ed.* **2018**, *57*, 16313–16317. [CrossRef]
88. Feyand, M.; Mugnaioli, E.; Vermoortele, F.; Bueken, B.; Dieterich, J.M.; Reimer, T.; Kolb, U.; De Vos, D.; Stock, N. Automated diffraction tomography for the structure elucidation of twinned, sub-micrometer crystals of a highly porous, catalytically active bismuth metal-organic framework. *Angew. Chem. Int. Ed.* **2012**, *51*, 10373–10376. [CrossRef]
89. Kolb, U.; Mugnaioli, E.; Gorelik, T.E. Automated electron diffraction tomography—A new tool for nano crystal structure analysis. *Cryst. Res. Technol.* **2011**, *46*, 542–554. [CrossRef]

90. Mintova, S.; Petit, S.; Zaarour, M.; Boullay, P.; Perez, O.; Brázda, P.; Eigner, V.; Palatinus, L.; Klementová, M. Hydrogen positions in single nanocrystals revealed by electron diffraction. *Science* **2017**, *355*, 166–169. [CrossRef]
91. Wan, W.; Sun, J.; Su, J.; Hovmöller, S.; Zou, X. Three-dimensional rotation electron diffraction: Software RED for automated data collection and data processing. *J. Appl. Crystallogr.* **2013**, *46*, 1863–1873. [CrossRef]
92. Walton, R.I.; Millange, F. In Situ Studies of the Crystallization of Metal-Organic Frameworks. In *The Chemistry of Metal-Organic Frameworks: Synthesis, Characterization, and Applications*; Wiley-VCH Verlag GmbH & Co. KGaA: Weinheim, Germany, 2016; pp. 729–764.
93. Heidenreich, N.; Rütt, U.; Suren, R.; Stock, N.; Inge, A.K.; Dippel, A.-C.; Köppen, M.; Beier, S. A multi-purpose reaction cell for the investigation of reactions under solvothermal conditions. *Rev. Sci. Instrum.* **2017**, *88*, 104102. [CrossRef] [PubMed]
94. Van Vleet, M.J.; Weng, T.; Li, X.; Schmidt, J.R. In Situ, Time-Resolved, and Mechanistic Studies of Metal-Organic Framework Nucleation and Growth. *Chem. Rev.* **2018**, *118*, 3681–3721. [CrossRef] [PubMed]
95. Feyand, M.; Näther, C.; Rothkirch, A.; Stock, N. Systematic and in situ energy dispersive X-ray diffraction investigations on the formation of lanthanide phosphonatobutanesulfonates: Ln(O_3P-C_4H_8-SO_3)(H_2O) (Ln = La-Gd). *Inorg. Chem.* **2010**, *49*, 11158–11163. [CrossRef]
96. Feyand, M.; Hübner, A.; Rothkirch, A.; Wragg, D.S.; Stock, N. Copper phosphonatoethanesulfonates: Temperature dependent in situ energy dispersive x-ray diffraction study and influence of the pH on the crystal structures. *Inorg. Chem.* **2012**, *51*, 12540–12547. [CrossRef] [PubMed]
97. Batzdorf, L.; Fischer, F.; Wilke, M.; Wenzel, K.J.; Emmerling, F. Direct in situ investigation of milling reactions using combined X-ray diffraction and Raman spectroscopy. *Angew. Chem. Int. Ed.* **2015**, *54*, 1799–1802. [CrossRef]
98. Rothenberg, G. *Catalysis: Concepts and Green Applications*; Wiley-VCH: Weinheim, Germany, 2017; ISBN 3527808884.
99. Schlögl, R. Heterogeneous catalysis. *Angew. Chem. Int. Ed.* **2015**, *54*, 3465–3520. [CrossRef]
100. Thomas, J.M.; Thomas, W.J. *Principles and Practice of Heterogeneous Catalysis*, 2nd ed.; Wiley: Hoboken, NJ, USA, 2015; ISBN 9783527314584.
101. White, R.J.; Luque, R.; Budarin, V.L.; Clark, J.H.; MacQuarrie, D.J. Supported metal nanoparticles on porous materials. Methods and applications. *Chem. Soc. Rev.* **2009**, *38*, 481–494. [CrossRef] [PubMed]
102. Astruc, D. (Ed.) *Nanoparticles and Catalysis*; Wiley-VCH Verlag GmbH & Co. KGaA: Weinheim, Germany, 2008; ISBN 9783527315727.
103. Hu, Z.; Zhao, D. Metal-organic frameworks with Lewis acidity: Synthesis, characterization, and catalytic applications. *CrystEngComm* **2017**, *19*, 4066–4081. [CrossRef]
104. Ranocchiari, M.; van Bokhoven, J.A. Catalysis by metal-organic frameworks: Fundamentals and opportunities. *Phys. Chem. Chem. Phys.* **2011**, *13*, 6388–6396. [CrossRef]
105. Clearfielda, A.; Thakurb, D.S. Zirconium and titanium phosphates as catalysts: A review. *Appl. Catal.* **1986**, *26*, 1–26. [CrossRef]
106. Wang, Z.; Heising, J.M.; Clearfield, A. Sulfonated microporous organic-inorganic hybrids as strong bronsted acids. *J. Am. Chem. Soc.* **2003**, *125*, 10375–10383. [CrossRef]
107. Donnadio, A.; Nocchetti, M.; Costantino, F.; Taddei, M.; Casciola, M.; Da Silva Lisboa, F.; Vivani, R. A layered mixed zirconium phosphate/phosphonate with exposed carboxylic and phosphonic groups: X-ray powder structure and proton conductivity properties. *Inorg. Chem.* **2014**, *53*, 13220–13226. [CrossRef] [PubMed]
108. Costantino, F.; Vivani, R.; Bastianini, M.; Ortolani, L.; Piermatti, O.; Nocchetti, M.; Vaccaro, L. Accessing stable zirconium carboxy-aminophosphonate nanosheets as support for highly active Pd nanoparticles. *Chem. Commun.* **2015**, *51*, 15990–15993. [CrossRef] [PubMed]
109. Bastianini, M.; Costantino, F.; Caporali, M.; Liguori, F.; Nocchetti, M.; Lavacchi, A. Robust Zirconium Phosphate–Phosphonate Nanosheets Containing Palladium Nanoparticles as Efficient Catalyst for Alkynes and Nitroarenes Hydrogenation Reactions. *ACS Appl. Nano Mater.* **2018**, *1*, 1750–1757. [CrossRef]
110. Nocchetti, M.; Giannoni, T.; Vaccaro, L.; Piermatti, O.; Kozell, V.; Vivani, R. Immobilized Palladium Nanoparticles on Zirconium Carboxy-Aminophosphonates Nanosheets as an Efficient Recoverable Heterogeneous Catalyst for Suzuki–Miyaura and Heck Coupling. *Catalysts* **2017**, *7*, 186. [CrossRef]

111. Ferlin, F.; Cappelletti, M.; Vivani, R.; Pica, M.; Piermatti, O.; Vaccaro, L. Au@zirconium-phosphonate nanoparticles as an effective catalytic system for the chemoselective and switchable reduction of nitroarenes. *Green Chem.* **2019**, *21*, 614–626. [CrossRef]
112. Pica, M.; Donnadio, A.; Capitani, D.; Vivani, R.; Troni, E.; Casciola, M. Advances in the chemistry of nanosized zirconium phosphates: A new mild and quick route to the synthesis of nanocrystals. *Inorg. Chem.* **2011**, *50*, 11623–11630. [CrossRef]
113. Mitchell, L.; Gonzalez-Santiago, B.; Mowat, J.P.S.; Gunn, M.E.; Williamson, P.; Acerbi, N.; Clarke, M.L.; Wright, P.A. Remarkable Lewis acid catalytic performance of the scandium trimesate metal organic framework MIL-100(Sc) for C–C and C=N bond-forming reactions. *Catal. Sci. Technol.* **2013**, *3*, 606–617. [CrossRef]
114. Beier, M.J.; Kleist, W.; Wharmby, M.T.; Kissner, R.; Kimmerle, B.; Wright, P.A.; Grunwaldt, J.D.; Baiker, A. Aerobic epoxidation of olefins catalyzed by the cobalt-based metal-organic framework STA-12(Co). *Chem. Eur. J.* **2012**, *18*, 887–898. [CrossRef] [PubMed]
115. Oschatz, M.; Antonietti, M. A search for selectivity to enable CO_2 capture with porous adsorbents. *Energy Environ. Sci.* **2018**, *11*, 57–70. [CrossRef]
116. Li, B.; Chen, B. Fine-Tuning Porous Metal-Organic Frameworks for Gas Separations at Will. *Chem* **2016**, *1*, 669–671. [CrossRef]
117. Hermer, N.; Wharmby, M.T.; Stock, N. Re-Determination of the Crystal Structure of MIL-91(Al). *Z. Anorg. Allg. Chem.* **2017**, *643*, 137–140. [CrossRef]
118. Miller, S.R.; Pearce, G.M.; Wright, P.A.; Bonino, F.; Chavan, S.; Bordiga, S.; Margiolaki, I.; Guillou, N.; Férey, G.; Bourrelly, S.; et al. Structural transformations and adsorption of fuel-related gases of a structurally responsive nickel phosphonate metal-organic framework, Ni-STA-12. *J. Am. Chem. Soc.* **2008**, *130*, 15967–15981. [CrossRef]
119. García, E.J.; Mowat, J.P.S.; Wright, P.A.; Pérez-Pellitero, J.; Jallut, C.; Pirngruber, G.D. Role of structure and chemistry in controlling separations of CO_2/CH_4 and $CO_2/CH_4/CO$ mixtures over honeycomb MOFs with coordinatively unsaturated metal sites. *J. Phys. Chem. C* **2012**, *116*, 26636–26648. [CrossRef]
120. *World Energy Outlook 2017—Executive Summary*; International Energy Agency: Paris, France, 2017.
121. Winter, M.; Brodd, R.J. What Are Batteries, Fuel Cells, and Supercapacitors? *Chem. Rev.* **2004**, *104*, 4245–4270. [CrossRef]
122. Bagotsky, V.S.; Skundin, A.M.; Volfkovich, Y.M. *Electrochemical Power Sources: Batteries, Fuel Cells, and Supercapacitors*; John Wiley & Sons, Inc.: Hoboken, NJ, USA, 2015; ISBN 9781118942857.
123. Peng, B.; Chen, J. Functional materials with high-efficiency energy storage and conversion for batteries and fuel cells. *Coord. Chem. Rev.* **2009**, *253*, 2805–2813. [CrossRef]
124. Kreuer, K.D. Proton conductivity: Materials and applications. *Chem. Mater.* **1996**, *8*, 610–641. [CrossRef]
125. Alberti, G.; Casciola, M. Solid state protonic conductors, present main applications and future prospects. *Solid State Ion.* **2001**, *145*, 3–16. [CrossRef]
126. Vivani, R.; Casciola, M.; Alberi, G.; Costantino, U.; Peraio, A. Proton conductivity of zirconium carboxy n-alkyl phosphonates with an α-layered structure. *Solid State Ion.* **1991**, *46*, 61–68. [CrossRef]
127. Alberti, G.; Casciola, M.; Palombari, R.; Peraio, A. Protonic conductivity of layered zirconium phosphonates containing –SO_3H groups. II. Ac conductivity of zirconium alkyl-sulphophenyl phosphonates in the range 100–200 °C, in the presence or absence of water vapour. *Solid State Ion.* **1992**, *58*, 339–344. [CrossRef]
128. Shimizu, G.K.H.; Taylor, J.M.; Dawson, K.W. Metal Organophosphonate Proton Conductors. In *Metal Phosphonate Chemistry*; Royal Society of Chemistry: Cambridge, UK, 2011; pp. 493–524.
129. Colodrero, R.M.P.; Olivera-Pastor, P.; Losilla, E.R.; Aranda, M.A.G.; Leon-Reina, L.; Papadaki, M.; McKinlay, A.C.; Morris, R.E.; Demadis, K.D.; Cabeza, A. Multifunctional lanthanum tetraphosphonates: Flexible, ultramicroporous and proton-conducting hybrid frameworks. *Dalton Trans.* **2012**, *41*, 4045–4051. [CrossRef]
130. Bazaga-García, M.; Colodrero, R.M.; Papadaki, M.; Garczarek, P.; Zoń, J.; Olivera-Pastor, P.; Losilla, E.R.; León-Reina, L.; Aranda, M.A.G.; Choquesillo-Lazarte, D.; et al. Guest Molecule-Responsive Functional Calcium Phosphonate Frameworks for Tuned Proton Conductivity. *J. Am. Chem. Soc.* **2014**, *136*, 5731–5739. [CrossRef]

131. Colodrero, R.M.P.; Papathanasiou, K.E.; Stavgianoudaki, N.; Olivera-Pastor, P.; Losilla, E.R.; Aranda, M.A.G.; León-Reina, L.; Sanz, J.; Sobrados, I.; Choquesillo-Lazarte, D.; et al. Multifunctional luminescent and proton-conducting lanthanide carboxyphosphonate open-framework hybrids exhibiting crystalline-to-amorphous-to-crystalline transformations. *Chem. Mater.* **2012**, *24*, 3780–3792. [CrossRef]
132. Colodrero, R.M.P.; Olivera-Pastor, P.; Losilla, E.R.; Hernández-Alonso, D.; Aranda, M.A.G.; Leon-Reina, L.; Rius, J.; Demadis, K.D.; Moreau, B.; Villemin, D.; et al. High Proton Conductivity in a Flexible, Cross-Linked, Ultramicroporous Magnesium Tetraphosphonate Hybrid Framework. *Inorg. Chem.* **2012**, *51*, 7689–7698. [CrossRef]
133. Bazaga-García, M.; Papadaki, M.; Colodrero, R.M.P.; Olivera-Pastor, P.; Losilla, E.R.; Nieto-Ortega, B.; Aranda, M.Á.G.; Choquesillo-Lazarte, D.; Cabeza, A.; Demadis, K.D. Tuning proton conductivity in alkali metal phosphonocarboxylates by cation size-induced and water-facilitated proton transfer pathways. *Chem. Mater.* **2015**, *27*, 424–435. [CrossRef]
134. Schmidt, S.; Sheptyakov, D.; Jumas, J.C.; Medarde, M.; Benedek, P.; Novák, P.; Sallard, S.; Villevieille, C. Lithium Iron Methylenediphosphonate: A Model Material for New Organic-Inorganic Hybrid Positive Electrode Materials for Li Ion Batteries. *Chem. Mater.* **2015**, *27*, 7889–7895. [CrossRef]
135. Schmidt, S.; Sallard, S.; Sheptyakov, D.; Nachtegaal, M.; Novák, P.; Villevieille, C. Fe and Co methylene diphosphonates as conversion materials for Li-ion batteries. *J. Power Sources* **2017**, *342*, 879–885. [CrossRef]
136. Schmidt, S.; Sallard, S.; Sheptyakov, D.; Novák, P.; Villevieille, C. Ligand influence in Li-ion battery hybrid active materials: Ni methylenediphosphonate: Vs. Ni dimethylamino methylenediphosphonate. *Chem. Commun.* **2017**, *53*, 5420–5423. [CrossRef]
137. Galezowska, J.; Gumienna-Kontecka, E. Phosphonates, their complexes and bio-applications: A spectrum of surprising diversity. *Coord. Chem. Rev.* **2012**, *256*, 105–124. [CrossRef]
138. Pazianas, M.; Abrahamsen, B.; Ferrari, S.; Russell, R.G.G. Eliminating the need for fasting with oral administration of bisphosphonates. *Ther. Clin. Risk Manag.* **2013**, *9*, 395–402. [CrossRef]
139. Papathanasiou, K.E.; Vassaki, M.; Spinthaki, A.; Alatzoglou, F.-E.G.; Tripodianos, E.; Turhanen, P.; Demadis, K.D. Phosphorus chemistry: From small molecules, to polymers, to pharmaceutical and industrial applications. *Pure Appl. Chem.* **2018**, *91*, 421–441. [CrossRef]

© 2019 by the authors. Licensee MDPI, Basel, Switzerland. This article is an open access article distributed under the terms and conditions of the Creative Commons Attribution (CC BY) license (http://creativecommons.org/licenses/by/4.0/).

Article

Mechanochemical Access to Elusive Metal Diphosphinate Coordination Polymer

Andrea Ienco [1,*], Giulia Tuci [1], Annalisa Guerri [2] and Ferdinando Costantino [3]

[1] Consiglio Nazionale delle Ricerche—Istituto di Chimica dei Composti OrganoMetallici (CNR-ICCOM) Via Madonna del Piano 10, Sesto Fiorentino, I-50019 Firenze, Italy; giulia.tuci@iccom.cnr.it
[2] Dipartimento di Chimica "Ugo Schiff", University of Florence, Via della Lastruccia 3, Sesto Fiorentino, I-50019 Firenze, Italy; annalisa.guerri@unifi.it
[3] Dipartimento di Chimica, Biologia e Biotecnologie, University of Perugia, Via Elce di Sotto 8, I-06123 Perugia, Italy; ferdinando.costantino@unipg.it
* Correspondence: andrea.ienco@iccom.cnr.it; Tel.: +39-055-5225-282

Received: 24 April 2019; Accepted: 25 May 2019; Published: 29 May 2019

Abstract: Several binary metal diphosphinate compounds (ML) have been reported for diphosphinate bonded by a single methylene fragment. In case of longer bridges, binary products are difficult to isolate in crystalline form. Here, using a solvent assisted mechano-chemistry synthesis, we report two new ML crystalline phases, one hydrated and one anhydrous. The hydrated phase is a 2D coordination polymer with an open framework structure. Its network displays a new topology for coordination polymers and metal organic frameworks. The thermal behavior of the two phases has been studied. Finally, the importance of the bridge length is discussed in view of known metal diphosphinate compounds.

Keywords: coordination polymers; diphosphinate; copper; MOF; mechanochemistry

1. Introduction

For diphosphinates connected by a long bridge, a trio is better than a duet. We recall that phosphinates are a class of ligands widely studied in the 1960s and 1970s of the twentieth century to build inorganic polymers [1]. Nowadays, after a few decades of forgetfulness, a new use has been found for phosphinates as flame retardants, by replacing brominate polymers banned by the European Community [2]. In the last few years, phosphinates have also been used as novel ligands for the synthesis of robust Metal-organic Frameworks (MOF). For instance, in 2018 Demel and co-workers used phenylene-1,4-bis(methylphosphinic acid) (PBPA) for the synthesis of porous iron(II) MOF [3]. In this study, we focus on a particular class of phosphinates, namely those with two phosphinate groups connected by an organic bridge. In the last 10 years, we have extensively studied P,P'-diphenylmethylene diphosphinic acid (H_2pcp), P,P'-diphenylethylene diphosphinic acid (H_2pc$_2$p) and P,P'-diphenyl-p-xylylene diphosphinic acid (H_2pxylp) [4,5] as shown in Scheme 1. Different diphosphinic acids were also reported [6,7].

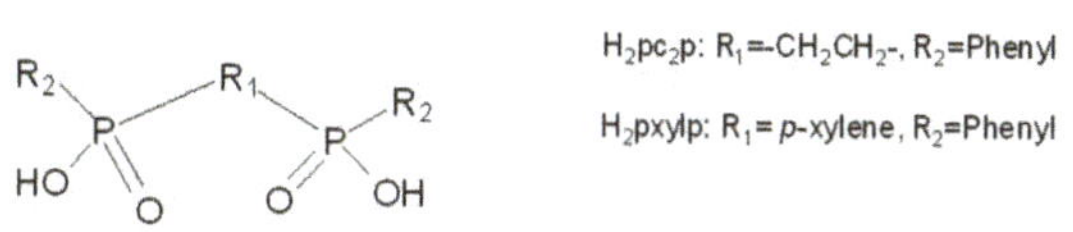

H_2pcp: R_1=-CH_2-, R_2=Phenyl
H_2pc$_2$p: R_1=-CH_2CH_2-, R_2=Phenyl
H_2pxylp: R_1= *p*-xylene, R_2=Phenyl

Scheme 1. Molecular drawing of bisphosphinate with different substituents.

In particular, the coordination ability of pcp towards alkaline earth metals [8–10], bivalent cations such tin [11] and lead [12] as well as several transition metal ions [11,13–17] are previously reported.

Depending on the metal and the use of ancillary ligands, mono- [14,18] and multinuclear complexes mboxciteB19-crystals-502202,B20-crystals-502202, 1D [21–23] or 2D [21,23] coordination polymers were obtained. The isolation of two isoreticular metal organic nano tubes (MONT) built by using copper(II) metal ion, pcp and two types of bi-pyridine was one of our most interesting results [24,25], also due to the different behavior of the two materials in water. [21] On the other hand, pc$_2$p and pxylp ligands were found to form 2D and 3D coordination polymers easily [26–28]. One of the latter, features porous channels able to reversibly adsorb methanol [26]. Under heating, different pseudo polymorph coordination polymers have been obtained upon removal of the coordinated water molecules [26–28].

We have already discussed the different coordination ability of pcp with respect to the other diphosphinate ligands with longer bridges [20]. From the reported structures, while pcp prefers to chelate the metal atoms, the other ligands connect different metal ions with their phosphinate units [26–28] and only when a co-ligand such as 2,2'-bipy leaves free, only the *cis* positions pc$_2$p is found to chelate a metal center [20]. We also noticed that binary metal pcp compounds are easily obtained, while the corresponding pc$_2$p systems are elusive, as the common synthetic routes employed (solution or hydrothermal methods) are not satisfactory. For instance, in the case of Ni, pc$_2$p, bipy systems, we obtained cationic 1D chains of $[Ni(H_2O)_4(bipy)]_n^{2+}$ and dianionic diphosphinate moieties not bonding to the metal center [28].

In this work, for the first time, we were able to isolate metal pc$_2$p coordination polymers, using Cu(II) as metal atom using a liquid assisted mechanochemical synthesis [29]. Depending on the amount of water used, a hydrous phase of formula $[[Cu_4(pc_2p)_4(H_2O)_6]\cdot 8(H_2O)]_n$, **1**, and Cu(pc$_2$p) anhydrous phase, **2**, were obtained. The X-ray analysis of **1** revealed a two-dimensional network with a potential porous structure, a unique characteristic in the family of the metal diphosphinate series. Topological analysis [30] was also carried out and it revealed a new topology type for **2** in the field of MOF and coordination polymers. In water **1** and **2** did not interconvert as shown by the thermal behavior of **1** and **2** studied by TGA and Temperature Dependent Powder Diffraction experiment (TDXD).

2. Materials and Methods

All reagents were analytical-grade commercial products and were used without further purification. The H$_2$pc$_2$p were prepared as previously described [31]. Elemental analyses (C, H, N) were performed with an EA 1108 CHNS-O automatic analyzer (Carlo Erba Instruments, Milan, Italy). XRD data was collected on a X'Pert PRO diffractometer (Panalytical, Almelo, the Netherlands) with Cu Kα radiation (λ = 1.5418 Å). Thermodiffractometric analysis were performed under air with an Anton Paar HTK 1200N hot chamber (Anton Paar, Graz, Austria). Thermogravimetric analyses (TGA) were performed on an EXSTAR Seiko 6200 analyzer (Seiko Instruments, Tokyo, Japan) under air (100 mL/min) at a heating rate of 10°C/min. The P/Cu ratio in 1 and 2 was measured using the X-ray (EDS) microanalysis system, Octane Elect Super Team Basic, (EDAX, AMETEK, Mahwah, NJ, USA) of a Gaia 3 (Tescan s.r.o, Brno, Czech Republic) FIB-SEM (Focused Ion Beam-Scanning Electron Microscope) Electron beam used for SEM imaging.

2.1. Synthesis of $[[Cu_4(pc_2p)_4(H_2O)_6]\cdot 8(H_2O)]_n$, 1

In an agate mortar, 1 mL of water was added to Cu acetate monohydrate (42 mg, 0.21 mmol). At the resulting slurry, H$_2$pc$_2$p (65 mg, 0.21 mmol) was added and grounded for 5 minutes. The product was washed with water and dried in air at room temperature. Well-formed blue crystals of **1** were obtained after several days by precipitation from a water solution (20 mL) of Cu acetate monohydrate (42 mg, 0.21 mmol) and H$_2$p$_2$pc (65 mg, 0.21 mmol). For the elemental analysis, the product dried in oven for 2 hours at 120 °C was used. The P/Cu ratio calculated using EDX is between 1.9 and 2.0 either for the dried and the hydrated product. Yield 63 mg, 86%. $C_{28}H_{26}Cu_2O_8P_4$ (741.50 gmol^{-1}): calcd. C 45.36, H 3.53; found: C 45.27, H 3.59.

2.2. Synthesis of Cupc$_2$p, 2

Cu acetate monohydrate (42 mg, 0.21 mmol) and H$_2$pc$_2$p (65 mg, 0.21 mmol) were ground with two drops of water for 5 minutes in an agate mortar. The product was washed with water and dried in air at room temperature. The P/Cu ratio calculated using EDX is between 1.9 and 2.0. Yield 55 mg, 88%. C$_{28}$H$_{26}$Cu$_2$O$_8$P$_4$ (741.50 gmol^{-1}): calcd. C 45.36, H 3.53; found: C 45.28, H 3.65.

2.3. X-Ray Structure Determination

A summary of the crystal data is given in Table 1. An Oxford Diffraction Excalibur 3 diffractometer (Oxford Diffraction Ltd., Abingdon, United Kingdom) equipped with Cu-Kα radiation and CCD area detector was used for data collection at 173K. The software CrysAlis CCD [32], CrysAlis RED [33] and ABSPACK [33] was employed for data collection, data reduction and absorption correction, respectively. Direct methods as coded in Sir97 [34] were used for structure solution. SHELXL program [35] was used for structure refinement on F^2 by full-matrix least squares techniques. All non-hydrogen atoms were refined anisotropically. Carbon bonded hydrogen atoms were introduced in calculated positions. The hydrogen atoms of water molecules (except the one bonded to OW5) were found in the Fourier map and refined using distance restraints. CCDC-1910569 contains the supplementary crystallographic data for this paper. These data can be obtained free of charge from the Cambridge Crystallographic Data Centre via www.ccdc.cam.ac.uk/data_request/cif.

Table 1. Crystal Data and Structure Refinement for 1.

Empirical formula	C56 H84 Cu4 O30 P8
Formula weight	1739.15
Temperature	153(2) K
Wavelength	1.5418 Å
Crystal system	Monoclinic
Space group	P 2/c
Unit cell dimensions	a = 10.4000(10) Å
	b = 10.355(2) Å
	c = 33.277(2) Å
	β = 99.033(8)°
Volume	3539.2(8) Å^3
Z	2
Density (calculated)	1.632 Mg/m^3
Absorption coefficient	3.793 mm^{-1}
F(000)	1792
Crystal size	0.16 × 0.15 × 0.12 mm^3
Theta range for data collection	4.269 to 72.397°
Index ranges	−12 < = h< = 12, −12 <= k<= 12, −41 <= l < = 34
Reflections collected	55473
Independent reflections	6660 [R$_{int}$ = 0.0430]
Completeness to θ = 26.06°	99.6%
Refinement method	Full-matrix least-squares on F^2
Data/restraints/parameters	6660/11/484
Goodness-of-fit on F^2	1.067
Final R indices [I > 2σ (I)]	R_1 = 0.0287, wR_2 = 0.0809
R indices (all data)	R_1 = 0.0337, wR_2 = 0.0847
Largest diff. peak and hole	0.451 and −0.621 e.Å^{-3}

3. Results and Discussion

Adding in an agata mortar 0.21 mmol of H$_2$pc$_2$p acid to a slurry of equivalent amount of copper(II) Acetate in 1 mL of water, a phase of formula [[Cu$_4$(pc$_2$p)$_4$(H$_2$O)$_6$]·8(H$_2$O)]$_n$, **1**, was obtained. The compound was recognized by the comparison of the X-ray powder diffraction (see Figure S1 in Supplementary Materials) with the calculated pattern of the structure resolved from single crystal grown in solution.

The structure of **1** resulted in a potentially porous open framework 2D coordination polymer and its formula is [[Cu$_4$(pc$_2$p)$_4$(H$_2$O)$_6$]·8(H$_2$O)]$_n$. The asymmetric unit contains two pc$_2$p ligands, three independent copper atoms and nine water molecules, four of them coordinated to copper ions and five as free solvent molecules. Selected distances and angles are reported in Table 2.

Table 2. Selected bond lengths (Å) and angles (°) for **2**. Symmetry transformations used to generate equivalent atoms: #1 x + 1, y, z; #2 x, y − 1,z; #3 −x, y, −z + 1/2; #4 −x + 1, y, −z + 1/2.

Cu1-O1	1.9221(15)	Cu3-O31	1.9790(15)	O3#2-Cu1-O7#1	88.89(6)
Cu1-O5	1.9258(14)	P2-O4	1.5115(16)	O8 -Cu2-O8#3	145.44(9)
Cu1-O3#2	1.9663(15)	P3-O5	1.5141(15)	O8-Cu2-O20	107.28(5)
Cu1-O7#1	1.9313(14)	P3-O6	1.5088(15)	O8-Cu2-O21	86.82(6)
Cu1-O20	2.4651(14)	P4-O7	1.5107(15)	O8#3-Cu2-O21	93.49(6)
Cu2-O8	1.9235(14)	P4-O8	1.5087(15)	O21-Cu2-O20	89.47(5)
Cu2-O20	2.203(2)	-	-	O21-Cu2-O21#3	178.95(9)
Cu2-O21	1.9938(15)			O6-Cu3-O6#4	179.06(9)
Cu3-O6	1.9293(14)	O1-Cu1-O3#2	178.36(7)	O6-Cu3-O30	90.47(4)
Cu3-O30	2.231(3)	O1-Cu1-O5	89.84(6)	O31-Cu3-O30	92.33(5)
P1-O1	1.5115(15)	O1-Cu1-O7#1	90.52(6)	O6-Cu3-O31	87.52(6)
P1-O2	1.5121(16)	O5-Cu1-O3#2	90.95(6)	O6-Cu3-O31#4	92.44(6)
P2-O3	1.5330(15)	O5-Cu1-O7#1	172.03(6)	O31-Cu3-O31#4	175.33(9)

Two metal atoms (Cu2 and Cu3) sit on a twofold axis, while the third is in general position. Also, four oxygen atoms of water molecules are on a twofold axis. As illustrated in Figure 1a, the first pc$_2$p ligand is bound to two different Cu1 atoms with O1 and O3 while the other two oxygen atoms are engaged in hydrogen bonds with five different water molecules. All the oxygen atoms of the second pc$_2$p ligand are connected with four different copper ions, namely Cu1, Cu1′, Cu2 and Cu3 (see Figure 1b). Before describing the coordination of copper atoms, it is useful to note that Cu1 metal and pc$_2$p ligands form a two dimensional square grid as shown in Figure 2a.

Figure 1c shows the coordination and the relation between Cu1 and Cu2 metal. Cu1 atom assumes a slightly distorted square pyramidal geometry. The basal positions are occupied by four oxygen atoms of four different pc$_2$p ligands, while the axial position is occupied by a water molecule with a distance Cu1-O20 of 2.4651(14)Å. The latter water is bridging between Cu1 and Cu2. The Cu2 atom is surrounded by three oxygen atoms of three water molecules and by two oxygen atoms of two different pc$_2$p ligands. In this case the geometry around the Cu2 atom is better described as trigonal bi-pyramidal. Finally, three oxygen atoms of three water molecules and two oxygen atoms of two different pc$_2$p ligand are around the Cu3 atom. A sixth water molecules is at 2.8852(6)Å as shown in Figure 1d. Interestingly, the three slightly different copper coordination in **1** is an example of how the Jahn Teller (or pseudo Jahn Teller) effect could influence the metal geometry in five coordinated complexes [36].

The pieces for building the two dimensional network are two square grids of Figure 2a on the *ab* plane and the strips formed by Cu1, Cu2 and Cu3 metal ions along *a* axis as shown in Figure 2b. The complete 2D layers are shown in Figure 2c and they have an approximate height of 16Å. The inner part presents rectangular channels of 7.7 × 9.2 Å^2 (excluding van der Waals radii) filled with water molecules. Excluding the solvent water molecules, the calculated void volume is 336Å^3 and it is more than the 5% of the unit cell volume. The layers are stacked together with the phenyl rings of the pc$_2$p ligands as shown in Figure 3c.

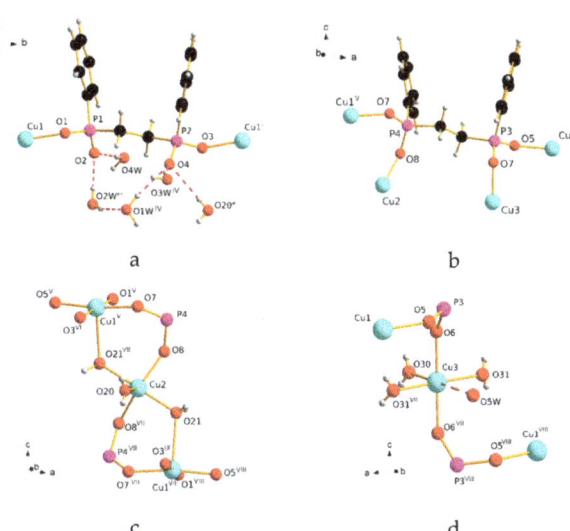

Figure 1. (a) Coordination mode of the first pc$_2$p ligand; (b) Coordination mode of the second pc$_2$p ligand; (c) Coordination of Cu1 and Cu2; (d) Coordination of Cu3. Color code: Cu turquoise, P purple, O red, C grey, H white. Symmetry transformations used to generate equivalent atoms: I = x, 1 + y, z; II = 1 + x, 1 + y, z; III = 1 − x, y, ½ − z; IV = 1 + x; y, z; V = x − 1, y, z; VI = x − 1, y − 1, z; VII = −x, y, ½ − z; VIII = 1 − x, y − 1, ½ − z.

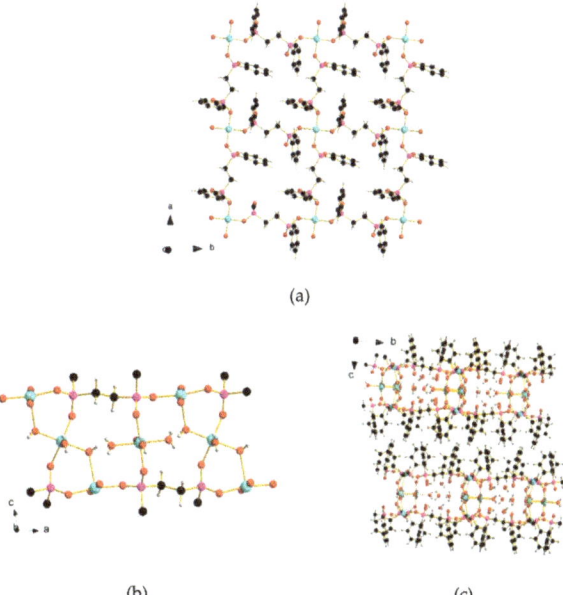

Figure 2. (a) Square grid formed by Cu1 and pc$_2$p ligands; (b) strips formed by by Cu1, Cu2 and Cu3 metal ions along *a* axis; (c) the packing of the 2D layers along the *c* axis. Color code: Cu turquoise, P purple, O red, C grey, H white.

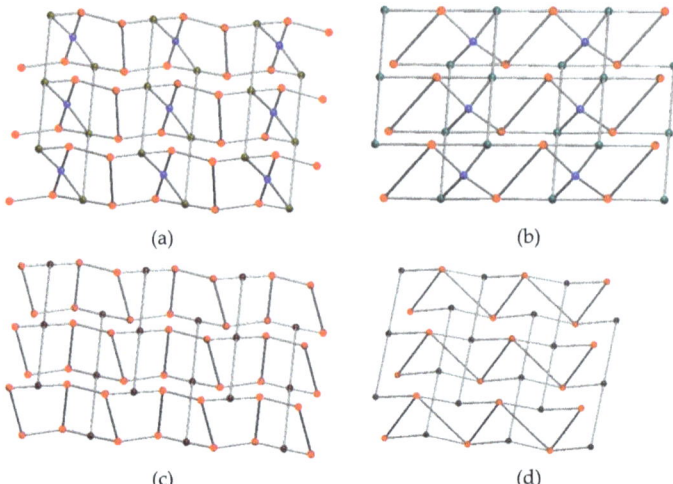

Figure 3. Schematic representation of; (**a**) 4 nodal network with two 3, one 4 and one 5 connected vertexes; (**b**) 3-nodal net with 4, 4, 5-connected vertex; (**c**) 4,4L62 and; (**d**) 3,4L156. The Cu1 node is represented in dark green, Cu3 in blue and P nodes in red.

The topological analysis was performed in order to better understand the connections in the 2D network and determine some correlation with known coordination polymers. [30,37–40] Looking at Figure 1c,d, Cu1 results in a 5-connected node, Cu2 in a 4-connected one, while Cu3 is only 2-connected and it does not contribute to the final topology. Also, the pc$_2$p ligand of P1 and P2 atoms is 2-connected (see Figure 1a). The other pc$_2$p could be considered in two different ways, in one case, in the so called standard representation, by two 3-connected nodes (P3 and P4 atoms) as shown in Figure 3a, in the other, the cluster reduction, the whole pc$_2$p ligand is substituted with a unique single 4-connected node (Figure 3b).

The network, represented in Figure 3a, is a 4 nodal network with two 3, one 4 and one 5 connected vertexes, while the one illustrated in Figure 3b is a 3-nodal net with a 4, 4, 5-connected vertex. The point symbols are $\{3.5.6^3.7^2.8^3\}2\{3.5.6\}2\{3^2.5^2.6^2\}\{5^2.8\}2$ and $\{3.4.5.6^6.8\}\ 2\{3.4^2.5.6^2\}\ 2\{3^2.4^2.5^2\}$ respectively. No network with this topology is present to date in the Topcryst topological database [41]. We also decided to not consider the bond between O20 and the Cu1, being longer than a normal Cu-O bond, in order to explore possible connections with other known networks. In this case, the Cu1 is a a 4-connected node, while the Cu2 is reduced to a 2-connected one. Again, the pc$_2$p ligand is considered formed by two 3-connected nodes in the standard representation (Figure 3c) and by a single 4 connected node in the cluster one (Figure 3d). With these choices, the network of Figure 3c is a 3-nodal net with two 3-connected and one 4 connected vertex, known as 3,4L156, while in the other case the 4,4L62 network is obtained as represented in Figure 3d. In the Topcryst database [41] two H-bonded molecular structures for 3,4L156 have been found (CCDC recodes: KEDBUS [42], EJIRIY [43]). Also, one H-bonded structure has the 4,4L62 topology (CCDC refcode ATEYII [44]). To the best of our knowledge, no covalent coordination polymer or MOF shows the same topology as **2**.

Using a different synthetic protocol, namely adding ater to a solid mixture of copper(II) Acetate and H$_2$pc$_2$p, a second phase, **2**, appeared in the powder X-ray diffraction pattern (see Figure 4d). Eventually, when just a small amount of water was used, pure **2** was synthetized. Compound **2** resulted in an anhydrous phase and it can be written as Cu(pc$_2$p). Unfortunately, due to the poor crystallinity of the product, it was not possible to index the diffraction pattern and we were unable to obtain a single crystal of **2**.

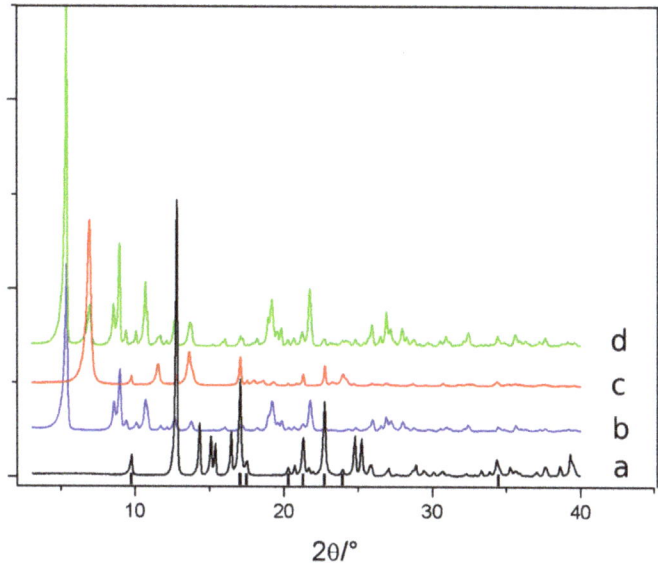

Figure 4. (a) Diffraction pattern of the Cu(CH$_3$COO)$_2$ H$_2$O and H$_2$pc$_2$p mixture, the main H$_2$pc$_2$p peaks in the bottom spectrum are indicated with black lines; (b) diffraction pattern of pure phase **1**; (c) diffraction pattern of pure phase **2** after the adding of few drops of water; (d) diffraction pattern of a mixture of **1** and **2** after the adding of 1ml of water.

The thermal behavior of the two phases was studied by thermogravimetric analysis (TGA) and temperature-dependent X-ray powder diffraction. For **1**, in the TGA under nitrogen (Figure 5a) the loss of solvent and coordinated water molecules of **1** is observed below 100 °C (14.5% calcd, 13.8% exp.). Above 200 °C, the degradation of the pc$_2$p ligand begins. The corresponding temperature-dependent X-ray powder diffraction patterns between 25 °C and 120 °C (Figure 5b) show a progressive meaningful shift of the first peak (0,0,2 of phase **1**) between 25 °C to 50 °C, indicating a shortening of the c axis. This could be interpreted as the shrinkage of the inorganic layers of **1**. In the pattern at 50 °C a new peak at around 5.9° 2θ appeared. At 70 °C, the transformation is complete as shown by the peak at 5.9° 2θ, which is attributable to an anhydrous **1**. Evidently, the latter has a different structural arrangement with respect to **2** and the loss of coordinated water results in a collapse of the inorganic layers. Phase **1** does not recover spontaneously at room temperature, but the crystallinity is restored when contacted with water. The TGA of **2** shows that the phase is stable up to 130 °C as shown in Figure 5c without any important weight loss, confirming the anhydrous nature of **2**. The temperature-dependent X-ray powder diffraction analysis between 25 °C and 120 °C shows no change in the diffraction pattern (see Figure S2 in Supplementary Materials). The total weight loss up to 800 °C is about 44% which is compatible with the thermal degradation of the organic part which occurs in two different steps (calc. 46.1% with the formula Cupc$_2$p).

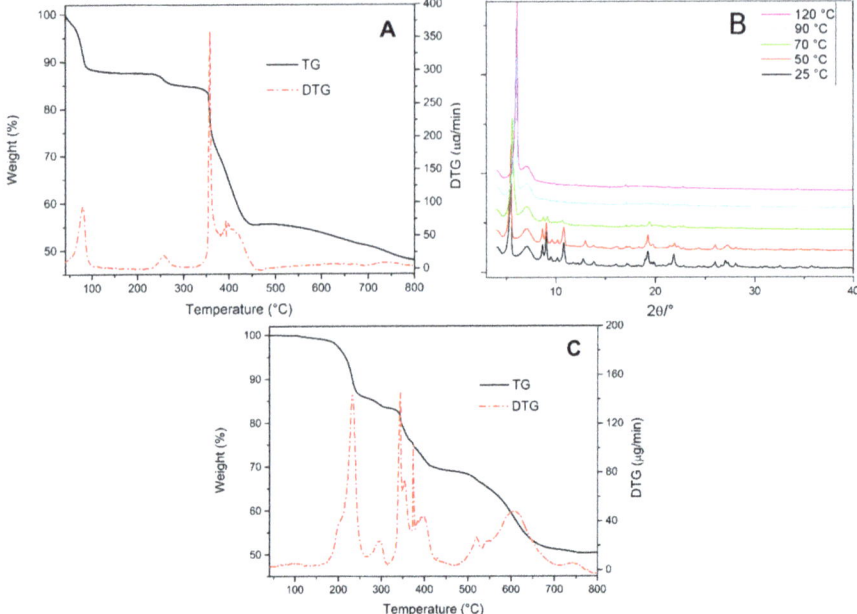

Figure 5. (**a**) TG/DTG curve of phase **1**; (**b**) temperature-dependent X-ray powder diffraction patterns for phase **1**, the peak between 6° and 8° 2θ is due to the diffraction of Kapton® polymer used as window in the Anton-Paar Hot Chamber; (**c**) TG/DTG curve of phase **2**.

Finally using CSD database surveys [45], we wanted to confirm the higher tendency of pcp for metal chelation compared to the pc₂p ligand. Only in three cases out of 28, pcp does not chelate to at least one metal center. The result is reversed for pc₂p (2 out of 7). The two examples are with a 2,2'-bipy co-ligand where they leave free only the cis positions around the metal center for the metal coordination of the pc₂p [20]. Another geometrical parameter is also investigated, namely the CPPC dihedral angle between the two phenyl rings bonded to the phosphorus atoms. For the 37 examples found, the average value is 28° for pcp with only three outliers with angles greater than 110°. Again, for pc₂p, the result is reversed. The mean value for the angle is 148° (15 examples) with only an outlier with an angle of 18°. So, the different behavior in metal coordination as well as in the ability of forming network and coordination polymers can be related to the different structural capability given by the nature of the bridge between the two phosphinate groups.

4. Conclusions

Using a solvent assisted mechano chemistry synthesis, two new binary phases between copper and pc₂p were been isolated. One of them was anhydrous material, while the other was found to be a peculiar 2D coordination polymer. Two square slabs were connected together by a copper metal center, forming potential voids filled by solvent water molecules. From the topological analysis, the 2D network revealed an unknown topology for coordination polymers, also when different kinds of network simplifications was considered. The thermal behavior of the two phases was studied. The collapse of the 2D network was confirmed, but the anhydrous phase obtained was different from the one directly obtained in the synthesis. A database survey on the known pcp and pc₂p structures demonstrated how the nature of the bridging chain between the phosphinate group is a key factor for the resulting structure.

Supplementary Materials: The following are available online at http://www.mdpi.com/2073-4352/9/6/283/s1, Figure S1: Comparison of the calculated and experimental X-ray pattern for **1**; Figure S2: Temperature-dependent X-ray powder diffraction patterns for phase **2**.

Author Contributions: Conceptualization, A.I. and F.C.; investigation, A.I., G.T., A.G., F.C.; writing—original draft preparation, A.I.; writing—review and editing, A.I., G.T., A.G., F.C.

Funding: This research received no external funding.

Acknowledgments: A.I. acknowledges Carlo Bartoli for his technical assistance. The authors thank the Italian National Research Council (CNR) microscopy facility "Ce.M.E.–Centro Microscopie Elettroniche Laura Bonzi" for the experiments on Gaia 3, instrument acquired thanks to "Ente Cassa di Risparmio di Firenze" Grant Number n. 2013.0878 and Regione Toscana POR FESR 2014-2020 for the project FELIX (Fotonica ed Elettronica Integrate per l'Industria), Grant Number 6455.

Conflicts of Interest: The authors declare no conflict of interest.

References

1. Vioux, A.; Le Bideau, J.; Mutin, P.H.; Leclercq, D. Hybrid Organic-Inorganic Materials Based on Organophosphorus Derivatives. In *New Aspects in Phosphorus Chemistry IV. Topics in Current Chemistry*; Majoral, J.P., Ed.; Springer: Berlin/Heidelberg, Germany, 2004; Volume 232, pp. 145–174.
2. Wendels, S.; Chavez, T.; Bonnet, M.; Salmeia, K.; Gaan, S. Recent developments in organophosphorus flame retardants containing P-C bond and their applications. *Materials* **2017**, *10*, 784. [CrossRef]
3. Hynek, J.; Brázda, P.; Rohlíček, J.; Londesborough, M.; Demel, J. Phosphinic Acid Based Linkers: Building Blocks in Metal–Organic Framework Chemistry. *Angew. Chem. Int. Ed.* **2018**, *57*, 5016–5019. [CrossRef]
4. Costantino, F.; Ienco, A.; Taddei, M. Hybrid Multifunctional Materials Based on Phosphonates, Phosphinates and Auxiliary Ligands. In *Tailored Organic-Inorganic Materials*; Brunet, E., Colon, J.L., Clearfield, A., Eds.; John Wiley & Sons: Hoboken, NJ, USA, 2015; pp. 193–244.
5. Costantino, F.; Ienco, A.; Taddei, M. The Influence of Non-covalent Interactions in the Structure and Dimensionality of Hybrid Compounds and Coordination Polymers in Non-Covalent Interactions. In *Synthesis and Design of New Compounds*; Maharramov, A.M., Mahmudov, K.T., Kopylovich, M.N., Pombeiro, A.J.L., Eds.; John Wiley & Sons: Hoboken, NJ, USA, 2016; pp. 163–184.
6. David, T.; Procházková, S.; Kotek, J.; Kubíček, V.; Hermann, P.; Lukeš, I. Aminoalkyl-1,1-bis(phosphinic acids): Stability, acid-base, and coordination properties. *Eur. J. Inorg. Chem.* **2014**, *2014*, 4357–4368. [CrossRef]
7. David, T.; Procházková, S.; Havlíčková, J.; Kotek, J.; Kubíček, V.; Hermann, P.; Lukeš, I. Methylene-bis[(aminomethyl)phosphinic acids]: Synthesis, acid-base and coordination properties. *Dalton Trans.* **2013**, *42*, 2414–2422. [CrossRef] [PubMed]
8. Cecconi, F.; Dominguez, S.; Masciocchi, N.; Midollini, S.; Sironi, A.; Vacca, A. Complexation of beryllium(II)ion by phosphinate ligands in aqueous solution. Synthesis and XRPD structure determination of Be[(PhPO$_2$)$_2$CH$_2$](H$_2$O)$_2$. *Inorg. Chem.* **2003**, *42*, 2350–2356. [CrossRef]
9. Midollini, S.; Lorenzo-Luis, P.; Orlandini, A. Inorganic-organic hybrid materials of p,p'-diphenylmethylenediphosphinic acid (H$_2$pcp) with magnesium and calcium ions: Synthesis and characterization of [Mg(Hpcp)$_2$], [Mg(Hpcp)$_2$(H$_2$O)$_4$], [Mg(pcp)(H$_2$O)$_3$](H$_2$O), [Ca(Hpcp)$_2$] and [Ca(pcp)(H$_2$O)] complexes. *Inorg. Chim. Acta* **2006**, *359*, 3275–3282. [CrossRef]
10. Costantino, F.; Gentili, P.; Guerri, A.; Ienco, A.; Midollini, S.; Oberhauser, W. Structural similarities in 1D coordination polymers of alkaline earth diphosphinates. *Inorg. Chim. Acta* **2012**, *391*, 150–157.
11. Beckmann, J.; Costantino, F.; Dakternieks, D.; Duthie, A.; Ienco, A.; Midollini, S.; Mitchell, C.; Orlandini, A.; Sorace, L. Inorganic-organic hybrids of the p,p'-diphenylmethylenediphosphinate, pcp^{2-}. Synthesis, characterization, and XRPD structures of [Sn(pcp)] and [Cu(pcp)]. *Inorg. Chem.* **2005**, *44*, 9416–9423.
12. Cecconi, F.; Ghilardi, C.; Midollini, S.; Orlandini, A. A new lead(II) inorganic-organic hybrid of the P,P'-diphenylmethylene-diphosphinate ligand: Synthesis and X-ray characterization of the [Pb(CH$_2$(P(Ph)O$_2$)$_2$)] complex. *Inorg. Chem. Commun.* **2003**, *6*, 546–548. [CrossRef]
13. Lyssenko, K.A.; Vologzhanina, A.V.; Torubaev, Y.V.; Nelyubina, Y.V. A comparative study of a mixed-ligand Copper(II) complex by the theory of atoms in molecules and the Voronoi tessellation. *Mendeleev Commun.* **2014**, *24*, 216–218. [CrossRef]

14. Ciattini, S.; Costantino, F.; Lorenzo-Luis, P.; Midollini, S.; Orlandini, A.; Vacca, A. Inorganic-organic hybrids formed by p,p'- diphenylmethylenediphosphinate, pcp^{2-}, with the Cu^{2+} ion. x-ray crystal structures of [Cu(pcp)H$_2$O$_2$]·H$_2$O and [Cu(pcp)bipy(H$_2$O)]. *Inorg. Chem.* **2005**, *44*, 4008–4016. [CrossRef]
15. Cecconi, F.; Dakternieks, D.; Duthie, A.; Ghilardi, C.; Gili, P.; Lorenzo-Luis, P.; Midollini, S.; Orlandini, A. Inorganic-organic hybrids of the p,p'-diphenylmethylenediphosphinate ligand with bivalent metals: A new 2D-layered phenylphosphinate zinc(II) complex. *J. Solid State Chem.* **2004**, *177*, 786–792. [CrossRef]
16. Berti, E.; Cecconi, F.; Ghilardi, C.; Midollini, S.; Orlandini, A.; Pitzalis, E. Isostructural organic-inorganic hybrids of P,P'-diphenyl-methylenediphosphinate (CH$_2$(P(Ph)O$_2$)$_2$)$^{2-}$ with divalent transition metals. *Inorg. Chem. Commun.* **2002**, *5*, 1041–1043.
17. Costantino, F.; Midollini, S.; Orlandini, A.; Sorace, L. Hydrothermal synthesis and structural characterization of a new 2D-layered vanadium diphosphinate: [VO(O$_2$(C$_6$H$_5$)PCH$_2$P(C$_6$H$_5$)O$_2$)]. *Inorg. Chem. Commun.* **2006**, *9*, 591–594. [CrossRef]
18. Midollini, S.; Orlandini, A. Hydrogen bonding in triamine copper(II) P,P'- Diphenyl methylene diphosphinate (pcp^{2-}) hybrids. Syntheses and crystal structures of [Cu(pcp)(2,2'-dipyridylamine)(H$_2$O)]·2H$_2$O and [Cu(pcp)(2,2':6',2''terpyridine)]·4 H$_2$O. *J. Coord. Chem.* **2006**, *59*, 1433–1442. [CrossRef]
19. Ienco, A.; Midollini, S.; Orlandini, A.; Costantino, F. Synthesis and structural characterization of a tetranuclear zinc(II) complex with P,P'-diphenylmethylenediphosphinate (pcp) and 2,2'-bipyridine (2,2'-bipy) ligands. *Z. Naturforsch. B Chem. Sci.* **2007**, *62*, 1476–1480. [CrossRef]
20. Costantino, F.; Ienco, A.; Midollini, S.; Orlandini, A.; Sorace, L.; Vacca, A. Copper(II) complexes with bridging diphosphinates—The effect of the elongation of the aliphatic chain on the structural arrangements around the metal centres. *Eur. J. Inorg. Chem.* **2008**, *2008*, 3046–3055. [CrossRef]
21. Taddei, M.; Ienco, A.; Costantino, F.; Guerri, A. Supramolecular interactions impacting on the water stability of tubular metal-organic frameworks. *RSC Adv.* **2013**, *3*, 26177–26183. [CrossRef]
22. Ienco, A.; Caporali, M.; Costantino, F.; Guerri, A.; Manca, G.; Moneti, S.; Peruzzini, M. The quest for hydrogen bond-based metal organic nanotubes (MONT). *J. Coord. Chem.* **2014**, *67*, 3863–3872. [CrossRef]
23. Costantino, F.; Midollini, S.; Orlandini, A. Cobalt(II) and nickel(II) coordination polymers constructed from P,P'-diphenylmethylenediphosphinic acid (H$_2$pcp) and 4,4'-bipyridine (bipy): Structural isomerism in [Co(pcp)(bipy)$_{0.5}$(H$_2$O)$_2$]. *Inorg. Chim. Acta* **2008**, *361*, 327–334. [CrossRef]
24. Bataille, T.; Costantino, F.; Lorenzo-Luis, P.; Midollini, S.; Orlandini, A. A new copper(II) tubelike metal-organic framework constructed from P,P'-diphenylmethylenediphosphinic acid and 4,4'-bipyridine: Synthesis, structure, and thermal behavior. *Inorg. Chim. Acta* **2008**, *361*, 9–15. [CrossRef]
25. Bataille, T.; Bracco, S.; Comotti, A.; Costantino, F.; Guerri, A.; Ienco, A.; Marmottini, F. Solvent dependent synthesis of micro-and nano-crystalline phosphinate based 1D tubular MOF: Structure and CO$_2$ adsorption selectivity. *CrystEngComm* **2012**, *14*, 7170–7173. [CrossRef]
26. Bataille, T.; Costantino, F.; Ienco, A.; Guerri, A.; Marmottini, F.; Midollini, S. A snapshot of a coordination polymer self-assembly process: The crystallization of a metastable 3D network followed by the spontaneous transformation in water to a 2D pseudopolymorphic phase. *Chem. Commun.* **2008**, *2008*, 6381–6383. [CrossRef] [PubMed]
27. Costantino, F.; Ienco, A.; Midollini, S. Different structural networks determined by variation of the ligand skeleton in copper(II) diphosphinate coordination polymers. *Cryst. Growth Des.* **2010**, *10*, 7–10. [CrossRef]
28. Guerri, A.; Taddei, M.; Bataille, T.; Moneti, S.; Boulon, M.; Sangregorio, C.; Costantino, F.; Ienco, A. Same Not the Same: Thermally Driven Transformation of Nickel Phosphinate-Bipyridine One-Dimensional Chains into Three-Dimensional Coordination Polymers. *Cryst. Growth Des.* **2018**, *18*, 2234–2242. [CrossRef]
29. Bowmaker, G.A. Solvent-assisted mechanochemistry. *Chem. Commun.* **2013**, *49*, 334–348.
30. Alexandrov, E.V.; Shevchenko, A.; Blatov, V.A. Topological databases: Why do we need them for design of coordination polymers? *Cryst. Growth Des.* **2019**. [CrossRef]
31. Garst, M.E. Alkylation of Phenyl Phosphinic Acid. *Synth. Commun.* **1979**, *9*, 261–266. [CrossRef]
32. *CrysAlisCCD*; Version1.171.33.41 (release06-05-2009 CrysAlis171.NET); Oxford Diffraction Ltd.: Abingdon, UK, 2009.
33. *CrysAlisRED*; Version1.171.33.41 (release06-05-2009 CrysAlis171.NET); Oxford Diffraction Ltd.: Abingdon, UK, 2009.

34. Altomare, A.; Burla, M.C.; Camalli, M.; Cascarano, G.L.; Giacovazzo, C.; Guagliardi, A.; Moliterni, A.G.G.; Polidori, G.; Spagna, R. SIR97: A new tool for crystal structure determination and refinement. *J. Appl. Crystallogr.* **1999**, *32*, 115–119. [CrossRef]
35. Sheldrick, G.M. A short history of SHELX. *Acta Crystallogr. Sect. A Found. Crystallogr.* **2008**, *64*, 112–122. [CrossRef]
36. Bacci, M. Jahn-Teller effect in five-coordinated copper(II) complexes. *Chem. Phys. Lett.* **1986**, *104*, 191–199. [CrossRef]
37. Bonneau, C.; O'Keeffe, M.; Proserpio, D.; Blatov, V.; Batten, S.; Bourne, S.; Lah, M.; Eon, J.; Hyde, S.; Wiggin, S.; et al. Deconstruction of Crystalline Networks into Underlying Nets: Relevance for Terminology Guidelines and Crystallographic Databases. *Cryst. Growth Des.* **2018**, *18*, 3411–3418. [CrossRef]
38. Blatov, V.A.; Shevchenko, A.P.; Proserpio, D.M. Applied Topological Analysis of Crystal Structures with the Program Package ToposPro. *Cryst. Growth Des.* **2014**, *14*, 3576–3586. [CrossRef]
39. Alexandrov, E.V.; Blatov, V.A.; Kochetkov, A.V.; Proserpio, D.M. Underlying nets in three-periodic coordination polymers: Topology, taxonomy and prediction from a computer-aided analysis of the Cambridge Structural Database. *CrystEngComm* **2011**, *13*, 3947–3958. [CrossRef]
40. Mitina, T.G.; Blatov V., A. Topology of 2-Periodic Coordination Networks: Toward Expert Systems in Crystal Design. *Cryst. Growth Des.* **2013**, *13*, 1655–1664. [CrossRef]
41. TOPCryst Topological Database. Available online: https://topcryst.com/ (accessed on 14 April 2019).
42. Sachdeva, N.; Dolzhenko, A.; Keung Chui, W. Regioselective synthesis of pyrimido[1,2-a][1,3,5]triazin-6-ones via reaction of 1-(6-oxo-1,6-dihydropyrimidin-2-yl)guanidines with triethylorthoacetate: Observation of an unexpected rearrangement. *Org. Biomol. Chem.* **2012**, *10*, 4586–4596. [CrossRef]
43. Wu, Y.-J.; He, H.; Sun, L.-Q.; Wu, D.; Gao, Q.; Li, H.-Y. Synthesis of fluorinated 1-(3-Morpholin-4-yl-phenyl)-Ethylamines. *Bioorg. Med. Chem. Lett.* **2003**, *13*, 1725–1728. [CrossRef]
44. Li, C.; An, C.; Li, X.; Gao, S.; Cui, C.; Sun, H.; Wang, B. Triazole and Dihydroimidazole Alkaloids from the Marine Sediment-Derived Fungus Penicillium paneum SD-44. *J. Nat. Prod.* **2011**, *74*, 1331–1334. [CrossRef]
45. Groom, C.R.; Bruno, I.J.; Lightfoot, M.P.; Ward, S.C. The Cambridge Structural Database. *Acta Cryst.* **2016**, *B72*, 171–179. [CrossRef]

© 2019 by the authors. Licensee MDPI, Basel, Switzerland. This article is an open access article distributed under the terms and conditions of the Creative Commons Attribution (CC BY) license (http://creativecommons.org/licenses/by/4.0/).

Article

Platonic Relationships in Metal Phosphonate Chemistry: Ionic Metal Phosphonates

Konstantinos Xanthopoulos [1], Zafeiria Anagnostou [1], Sophocles Chalkiadakis [1], Duane Choquesillo-Lazarte [2], Gellert Mezei [3], Jan K. Zaręba [4], Jerzy Zoń [5] and Konstantinos D. Demadis [1,*]

[1] Crystal Engineering, Growth and Design Laboratory, Department of Chemistry, University of Crete, Voutes Campus, GR-71003 Crete, Greece; kostas.sharma@gmail.com (K.X.); zaphiri@gmail.com (Z.A.); sophocleschalk@gmail.com (S.C.)
[2] Laboratorio de Estudios Cristalográficos, IACT, CSIC-Universidad de Granada, 18100 Granada, Spain; duane.choquesillo@csic.es
[3] Department of Chemistry, Western Michigan University, Kalamazoo, MI 49008-5413, USA; gellert.mezei@wmich.edu
[4] Advanced Materials Engineering and Modeling Group, Wroclaw University of Science and Technology, Wyb. Wyspiańskiego 27, 50-370 Wrocław, Poland; jan.zareba@pwr.edu.pl
[5] Department of Thermodynamics, Theory of Machine and Thermal Systems, Faculty of Mechanical and Power Engineering, Wroclaw University of Science and Technology, Wyb. Wyspiańskiego 27, 50-370 Wrocław, Poland; jerzy.zon@pwr.edu.pl
* Correspondence: demadis@uoc.gr; Tel.: +30-2810-545051

Received: 15 May 2019; Accepted: 7 June 2019; Published: 11 June 2019

Abstract: Phosphonate ligands demonstrate strong affinity for metal ions. However, there are several cases where the phosphonate is found non-coordinated to the metal ion. Such compounds could be characterized as salts, since the interactions involved are ionic and hydrogen bonding. In this paper we explore a number of such examples, using divalent metal ions (Mg^{2+}, Ca^{2+}, Sr^{2+} and Ni^{2+}) and the phosphonic acids: *p*-aminobenzylphosphonic acid (H_2PABPA), tetramethylenediamine-*tetrakis*(methylenephosphonic acid) (H_8TDTMP), and 1,2-ethylenediphosphonic acid (H_4EDPA). The compounds isolated and structurally characterized are $[Mg(H_2O)_6]·[HPABPA]_2·6H_2O$, $[Ca(H_2O)_8]·[HPABPA]_2$, $[Sr(H_2O)_8]·[HPABPA]_2$, $[Mg(H_2O)_6]·[H_6TDTMP]$, and $[Ni(H_2O)_6]·[H_2EDPA]·H_2O$. Also, the coordination polymer $\{[Ni(4,4'-bpy)(H_2O)_4]·[H_2EDPA]·H_2O\}_n$ was synthesized and characterized, which contains a bridging 4,4'-bipyridine (4,4'-bpy) ligand forming an infinite chain with the Ni^{2+} cations. All these compounds contain the phosphonate anion as the counterion to charge balance the cationic charge originating from the metal cation.

Keywords: metal phosphonate; ionic compounds; phosphonic acids; organic salts

1. Introduction

The area of metal phosphonate chemistry has seen impressive growth in the last decade [1]. Metal phosphonate compounds are known for their complex and diverse structural motifs [2] and their unique structural topologies [3]. They are also candidates for use in versatile application areas such as proton conductivity [4], gas storage [5], pharmacology [6], ion-exchange [7], catalysis [8] and corrosion inhibition [9].

The phosphonate group, either mono-deprotonated ($-PO_3H^-$), or doubly-deprotonated ($-PO_3^{2-}$), demonstrates high affinity for metal ions [10]. Because of this, and the stability of the metal phosphonate coordination bonds, hybrid metal organic materials with non-coordinated phosphonate groups (Ionic

Metal-Phosphonates), are relatively scarce. However, notable examples are found in the literature, and are briefly presented below.

The hydrated salt [Co(NH$_3$)$_6$](O$_3$P(C$_6$H$_4$)$_2$COO)·4H$_2$O was crystallized from an aqueous ammonia solution of 4-phosphono-biphenyl-4'-carboxylic acid [11]. In a similar approach, 4-phosphonobenzoic acid was crystallized with hexaaquacobalt(II) [12]. The compound [Cd$_2$Cl$_4$(H$_2$O)$_6$]$_{0.5}$[H$_4$L] [L = 2-morpholinoethyliminobis(methylenephosphonic acid)] contains a Cd chlorodimer as the cationic moiety [13]. The salt [Ni(phen)$_3$](H$_7$L$_2$)$_{0.5}$(H$_5$L$_2$)$_{0.5}$·2H$_2$O (phen = 1,10-phenanthroline, L = hydroxyethylidene-1,1-diphosphonic acid) showed two forms of the ligand in the same structure, with different degrees of deprotonation [14]. A series of divalent transition metal phosphonates containing hydrogen-bonded layers of phenylphosphonate anions, namely [M(phen)$_3$]·C$_6$H$_5$PO$_3$·11H$_2$O [M = Co^{2+}, Ni^{2+}, Cu^{2+}] and [Cd(phen)$_3$]·C$_6$H$_5$PO$_3$H·Cl·7H$_2$O were synthesized and structurally characterized by single-crystal X-ray diffraction methods [15]. The synthesis and crystal structures of five new analogues of the supramolecular copper(II) arylphosphonate [Cu(phen)$_2$Cl][(C$_6$H$_5$PO$_3$H$_2$)(HO$_3$PC$_6$H$_5$)] were reported. The structures contain substituted phenylphosphonic acids, and are of the general formula [Cu(phen)$_2$Cl][(XPO$_3$H$_2$)(HO$_3$PX)]·Z, where X = o-CH$_3$(C$_6$H$_4$); X = p-CH$_3$(C$_6$H$_4$), Z = H$_2$O·2CH$_3$CH$_2$OH; X = o-NO$_2$(C$_6$H$_4$), m-NO$_2$(C$_6$H$_4$); X = m-NO$_2$(C$_6$H$_4$); X = C$_{10}$H$_7$ [16]. The salt [Mg$_{1.5}$(H$_2$O)$_9$]·(L-H$_2$)$_{1.5}$·6H$_2$O (L-H$_2$ = O$_3$PCH$_2$N(H)(C$_4$H$_8$)N(H)CH$_2$PO$_3$, N,N'-piperazine-*bis*(methylenephosphonic acid) was hydrothermally synthesized and structurally characterized [17]. The crystal structure of the zinc hexaaqua compound [Zn(H$_2$O)$_6$][TDTMP] (TDTMP = tetramethylenediamine-*tetrakis*(methylenephosphonate)) was reported [18]. Alkaline-earth metal phosphates containing nitrogen-donor ligands were prepared by the reaction of alkaline-earth metal acetates M(OAc)$_2$·xH$_2$O (M = Mg, Ca, Sr, Ba) with 2,6-diisopropylphenylphosphate (dippH$_2$). Interaction of strontium or barium acetate with dippH$_2$ in methanol at room temperature leads to the isolation of ionic phosphates [{M$_2$(μ-H$_2$O)$_4$(H$_2$O)$_{10}$}{dipp}$_2$]·4L [M = Sr, L = CH$_3$OH; M = Ba, L = H$_2$O] [19]. The structure of the compound bis(hydrogen diphenylmethylphosphonato)-magnesium octahydrate contains polar and non-polar layers. The polar layers contain Mg(H$_2$O)$_6^{2+}$ ions, water of hydration and the phosphonate O atoms, and the non-polar layers contain benzhydryl groups. Two-dimensional hydrogen-bonding networks link Mg(H$_2$O)$_6^{2+}$ and the water of hydration to the phosphonate O atoms [20]. Three 1D nickel coordination polymers based on P,P'-diphenylethylenediphosphinic acid and three different bis-pyridine co-ligands, namely 4,4'-bipyridine, 1,2-bis(4-pyridyl)ethane and 1,2-bis(4-pyridyl)ethane, were prepared in mild hydrothermal conditions. They all contained 1D "Ni-bipyridyl" chains, with the phosphinate playing the role of the counterion [21]. The synthesis and crystal structure of Mg(2-AEPH)$_2$·8H$_2$O (2-AEPH = 2-aminoethylphosphonate) were reported [22]. The compound is built from octahedral magnesium hexaaqua dications, uncoordinated 2-AEPH$^-$ anions, and interstitial water molecules. The latter have no metal contacts but are engaged in hydrogen bonding, in which the water molecules, the (protonated) amino functions and the O-acceptor sites of the anions participate. The structures of the alkaline earth metal (Mg, Ca, Sr and Ba) complexes with (4-aminophenyl)arsonic acid (p-arsanilic acid) have been determined [23]. Of these, only the Mg analog, hexaaquamagnesium bis[hydrogen (4-aminophenyl)arsonate] tetrahydrate, [Mg(H$_2$O)$_6$](C$_6$H$_7$AsNO$_3$)·4H$_2$O, is ionic. The octahedral [Mg(H$_2$O)$_6$]$^{2+}$ cation, the two hydrogen p-arsanilate anions and the four water molecules of solvation form a three-dimensional network structure through inter-species O—H and N—H hydrogen-bonding interactions with water and arsonate O-atom and amine N-atom acceptors.

In this study we present the synthesis and structural characterization of five novel ionic metal phosphonates containing divalent metal ions, such as alkaline-earth metals (M = Mg^{2+}, Ca^{2+}, Sr^{2+}) and the 3D transition metal Ni^{2+}. The alkaline-earth metal compounds contain the aromatic amino-phosphonate ligand p-aminobenzylphosphonic acid (H$_2$PABPA), or the tetraphosphonic acid ligand tetramethylenediamine-*tetrakis*(methylenephosphonic acid) (H$_8$TDTMP) (when M = Mg^{2+}). The two Ni^{2+}-containing compounds both contain the ligand 1,2-ethylenediphosphonic acid (EDPA), but one of them also contains the bridging co-ligand 4,4′-bipyridine. The latter is a Ni-4,4′-bipyridine coordination polymer having the EDPA as the counterion. Schematic structures of the ligand used herein are presented in Figure 1.

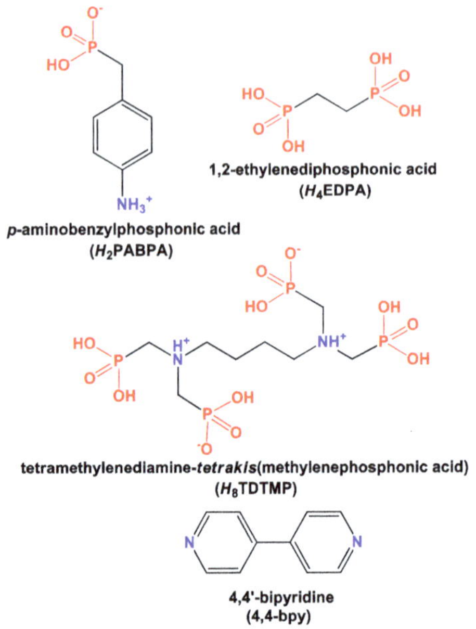

Figure 1. Phosphonic acids and N-heterocyclic ligand used in this paper. Phosphonato groups are highlighted in red, and N groups in blue. Amine-containing ligands are shown in their zwitterionic form.

2. Materials and Methods

2.1. Instrumentation

^{1}H, ^{31}P{^{1}H} NMR spectra were recorded on a Jeol JNM-ECZ 400S Research FT NMR spectrometer (JEOL Ltd., Tokyo, Japan) operating at 400 MHz and 161.7 MHz for ^{1}H and ^{31}P nuclei, respectively. ATR-IR spectra were collected on a Thermo-Electron NICOLET 6700 FTIR optical spectrometer (Thermo Fisher Scientific, Waltham, MA, USA). Elemental analyses (C, H, N) were measured on a Perkin–Elmer 2400 analyzer (Perkin–Elmer, Waltham, MA, USA). Thermogravimetric analysis (TGA) data were recorded on an SDT-Q600 analyzer from TA instruments (TA instruments, New Castle, DE, USA). The temperature varied from RT to 900 °C at a heating rate of 10 °C·min^{-1}. Measurements were carried out on samples in open platinum crucibles under air flow.

2.2. General

Starting materials were of reagent grade purity, obtained from commercial sources, and used without further purification. Deionized (DI) water was purified by a cation-exchange column.

2.3. Synthesis of 4-(bromomethyl)nitrobenzene

A 100 mL flask was charged with 4-(hydroxymethyl)nitrobenzene (5.00 g, 0.0326 mol), and 30% HBr solution in glacial acetic acid (35 mL). The obtained mixture was stirred at gentle reflux for 2 h. After cooling down, 48% HBr solution in water (20 mL) was added in order to precipitate as much crude 4-(bromomethyl)nitrobenzene as possible. The obtained brownish solid was filtered out, dissolved in chloroform and passed through a thick plug of finely ground K$_2$CO$_3$, to remove acidic impurities and residual water. After solvent evaporation, crude 4-(bromomethyl)nitrobenzene was crystallized from hexanes affording light beige crystals. Yield: 6.41 g (91%). Spectral data match those previously reported [24]. ^{1}H NMR (400 MHz, CDCl$_3$): δ 8.20 (d, $^3J_{HH}$ = 8.4 Hz, 2H), 7.55 (d, $^3J_{HH}$= 8.4 Hz, 2H), 4.51 (s, 2H).

2.4. Synthesis of 4-(diethoxyphosphorylmethyl)nitrobenzene

A three-neck 50 mL flask fitted with trap and gas inlet was charged with 4-(bromomethyl)nitrobenzene (5.00 g, 0.0231 mol), and triethyl phosphite (10 mL, 0.0583 mol). The mixture was heated to 120 °C under constant flow of nitrogen. Ethyl bromide, which is a coproduct of the Arbuzov reaction, was collected in the trap. After two hours of reaction the volatiles were distilled off under reduced pressure at the same temperature. Dark orange-colored crude product was subjected to column chromatography (silica gel, eluent: ethyl acetate, R_f = 0.40) yielding yellowish oil. Yield: 5.63 g (89%). Spectral data matched those previously reported [25]. ^{1}H NMR (400 MHz, CDCl$_3$): δ 8.24 (d, $^3J_{HH}$ = 8.7 Hz, 2H), 7.54 (dd, $^3J_{HH}$ = 8.7 Hz, $^4J_{PH}$ = 2.5 Hz, 2H), 3.93 (m, 4H), 3.30 (d, $^2J_{PH}$ = 22.4 Hz, 2H), 1.32 (t, $^3J_{HH}$ = 7.0 Hz, 6H). ^{31}P{^{1}H} NMR (161.7 MHz, CDCl$_3$): δ 25.10 (s).

2.5. Synthesis of 4-(diethoxyphosphorylmethyl)aniline

This procedure is an adaptation of a reduction protocol employed for the synthesis of 4-(diethoxyphosphoryl)aniline [26]. A 250 mL flask was charged with 4-(diethoxyphosphorylmethyl)nitrobenzene (4.80 g, 0.0175 mol), triethylamine (50 mL, 0.358 mol), and 100 mg 10%Pd/C catalyst. The reaction mixture was preheated to 50 °C and 90% Formic acid (8 mL, 0.18 mol) was added in small portions. Note that the reaction is strongly exothermic, and that hydrogen gas is evolved. After addition of all formic acid the reflux was maintained for 30 min, then the reaction mixture was cooled down, and evaporated. The crude product was dissolved in chloroform (100 mL), extracted with water (2 × 50 mL), and brine (50 mL). The organic phase was collected, dried with K$_2$CO$_3$, and evaporated. Column chromatography (silica gel, eluent: ethyl acetate, R_f = 0.22) provided pure 4-(diethoxyphosphorylmethyl)aniline as a light yellow oil that crystallizes upon standing. Yield: 3.35 g (79%). Spectral data match those previously reported [27]. ^{1}H NMR (400 MHz, CDCl$_3$) δ 7.05 (dd, $^3J_{HH}$ = 8.6 Hz, $^4J_{PH}$= 2.6 Hz, 2H), 6.61 (d, $^3J_{HH}$ = 7.9 Hz, 2H), 4.05–3.88 (m, 2H), 3.61 (br. s, 2H), 3.02 (d, $^2J_{PH}$ = 20.9 Hz, 4H), 1.22 (td, $^3J_{HH}$= 7.1, $^4J_{PH}$ = 0.5 Hz, 6H). ^{31}P{^{1}H} NMR (161.7 MHz, CDCl$_3$): δ 27.83 (s).

2.6. Synthesis of 4-(dihydroxyphosphorylmethyl)aniline (H$_2$PABPA)

4-(Diethoxyphosphorylmethyl)aniline (3.20 g, 0.0131 mol), concentrated hydrochloric acid (25 mL) and water (25 mL) were mixed together in a 100 mL flask. The obtained mixture was heated at reflux for 20 h. After hydrolysis the reaction mixture was evaporated to dryness. Crude product was boiled with ethanol (30 mL) for 30 min, cooled down and left overnight. A beige precipitate was filtered out and dried, providing pure 4-(dihydroxyphosphorylmethyl)aniline. Yield: 2.19 g (89%). ^{1}H NMR (400 MHz, D$_2$O + D$_2$SO$_4$) δ 7.16 (dd, $^3J_{HH}$ = 8.8 Hz, $^4J_{PH}$= 2.6 Hz, 2H), 7.09 (dd, $^3J_{HH}$ = 8.8 Hz, $^5J_{PH}$= 0.8 Hz, 2H), 2.99 (d, $^2J_{PH}$ = 21.5 Hz, 2H). ^{31}P{^{1}H} NMR (161.7 MHz, D$_2$O + D$_2$SO$_4$): δ 25.48 (s).

2.7. Synthesis of [Mg(H$_2$O)$_6$]·[HPABPA]$_2$·6H$_2$O (Mg-PABPA)

Solid H$_2$PABPA (132 mg, 0.7 mmol) was added to 140 mL of deionized water in a 250 mL glass beaker. A small quantity of 2 M NaOH solution was added until the pH of the solution reached the value of ~12.0. The solution was then heated to 80 °C and stirred vigorously until all the solid H$_2$PABPA dissolved. The solution was left to reach room temperature and then a small amount of a 2 M HCl solution was added until the pH dropped to 7.7. In a separate glass beaker, a quantity of MgCl$_2$·6H$_2$O (70 mg, 0.35 mmol) was dissolved in 70 mL of DI water. The two solutions were finally transferred to a polypropylene vessel and mixed under vigorous stirring. The pH of the resulting solution was adjusted to 7.0. This clear, colorless solution was left to stand at room temperature for

~2 weeks and product crystallization was achieved via slow evaporation of the solvent. The dark brown crystals were isolated by filtration, washed with small amounts of DI water, and left to air-dry. Yield: 55 mg (35%). Elemental analysis for [Mg(H$_2$O)$_6$]·[HPABPA]$_2$·6H$_2$O, MW 612.74. Calculated (Found) %: C 27.42 (27.39), H 6.85 (6.04), N 4.57 (4.60).

2.8. Synthesis of [Ca(H$_2$O)$_8$]·[HPABPA]$_2$ (Ca-PABPA)

The same method as for [Mg(H$_2$O)$_6$]·[HPABPA]$_2$·6H$_2$O was used for this synthesis. The reactants used were: H$_2$PABPA (132 mg, 0.7 mmol) and CaCl$_2$·2H$_2$O (50 mg, 0.35 mmol). Yield: 66 mg (34%). Elemental analysis for [Ca(H$_2$O)$_8$]·[HPABPA]$_2$, MW 556.45. Calculated (Found)%: C 29.39 (30.19), H 5.97 (6.11), N 4.92 (5.03).

2.9. Synthesis of [Sr(H$_2$O)$_8$]·[HPABPA]$_2$ (Sr-PABPA)

The same method as for [Mg(H$_2$O)$_6$]·[HPABPA]$_2$·6H$_2$O was used for this synthesis. The reactants used were: H$_2$PABPA (132 mg, 0.7 mmol) and SrCl$_2$·6H$_2$O (90 mg, 0.35 mmol). Yield: 70 mg (33%). Elemental analysis for [Sr(H$_2$O)$_8$]·[HPABPA]$_2$, MW 603.99. Calculated (Found)%: C 27.82 (25.87), H 5.63 (5.13), N 4.64 (4.35).

2.10. Synthesis of [Ni(H$_2$O)$_6$]·[H$_2$EDPA]·H$_2$O (Ni-EDPA)

Ni(NO$_3$)$_2$·6H$_2$O (290 mg, 1.0 mmol) and H$_4$EDPA (196 mg, 1.0 mmol) were simultaneously dissolved in DI water (100 mL) in a polypropylene vessel under vigorous stirring and the pH of the solution was adjusted to 4.0 with a small quantity of a 2 M NaOH solution. This clear, greenish solution was left to stand at room temperature and product crystallization was achieved via slow evaporation of the solvent. Emerald green crystals formed after 15 days and were filtered and washed with a small quantity of DI water. Yield: 217 mg (78%). Elemental analysis for [Ni(H$_2$O)$_6$]·[H$_2$EDPA]·H$_2$O, MW 390.84. Calculated (Found)%: C 6.14 (5.77), H 5.63 (5.53), N 0 (0.05).

2.11. Synthesis of {[Ni(4,4'-bpy)(H$_2$O)$_4$]·[H$_2$EDPA]·H$_2$O}$_n$ (Ni-bpy-EDPA)

Ni(NO$_3$)$_2$·6H$_2$O (290 mg, 1.0 mmol), H$_4$EDPA (196 mg, 1.0 mmol), and 4,4'-bpy (156 mg, 1.0 mmol) were simultaneously dissolved in DI water (100 mL) in a polypropylene (PP) vessel under vigorous stirring and the pH of the solution was adjusted to 5.0 with a small quantity of a 2 M NaOH solution. This clear, greenish solution was left to stand at room temperature and product crystallization was achieved via slow evaporation of the solvent. Emerald green crystals formed after 15 days and were filtered and washed with a small quantity of DI water. Yield: 219 mg (45%). Elemental analysis for [Ni(4,4'-bpy)(H$_2$O)$_4$]·[H$_2$EDPA]·H$_2$O, MW 511.00. Calculated (Found)%: C 28.18 (28.42), H 5.09 (5.73), N 5.48 (6.13).

2.12. Synthesis of [Mg(H$_2$O)$_6$]·[H$_6$TDTMP] (Mg-TDTMP)

In a PP vessel solid H$_8$TDTMP acid (46 mg, 0.1 mmol) was mixed with MgCl$_2$·6H$_2$O (20 mg, 0.1 mmol) in 10 mL DI water under continuous stirring until complete dissolution of the solids. The pH was adjusted to ~ 3.0. The clear colorless solution was then left to stand at ambient conditions for partial solvent evaporation. A crystalline precipitate was obtained after 15 days, isolated by filtration, washed with DI water and left to dry under air. Yield 36 mg (60%). Elemental analysis for [Mg(H$_2$O)$_6$]·[H$_6$TDTMP], MW 594.56. Calculated (Found)%: C 16.15 (16.31), H 5.72 (5.66), N 4.71 4.54).

2.13. Crystal Data Collection And Refinement

Crystals for measurements were handled under inert conditions. They were immersed in perfluoropolyether as a protecting oil for manipulation. Suitable single crystals were mounted on MiTeGen Micromounts™ and subsequently used for data collection. X-ray diffraction data for **Ni-bpy-EDPA** and **Mg-TDTMP** were collected at room temperature from a single-crystal mounted atop a glass fiber with cyanoacrylate glue, using a Bruker SMART APEX II diffractometer with graphite-monochromated Mo-Kα radiation. Data for **Mg-PABPA, Ca-PABPA, Sr-PABPA** and **Ni-EDPA** were collected with a Bruker D8 Venture diffractometer. The data were processed with the APEX3 suite [28]. The structures were solved by direct methods [29], which revealed the position of all non-hydrogen atoms. These atoms were refined on F^2 by a full-matrix least squares procedure using anisotropic displacement parameters [30]. All hydrogen atoms were located in difference Fourier maps and included as fixed contributions riding on attached atoms with isotropic thermal displacement parameters 1.2 or 1.5 times those of the respective atom. Crystallographic data for the reported structures have been deposited with the Cambridge Crystallographic Data Center as supplementary publication no. CCDC 1914866 – 1914868 and 1914870-1914872. Additional crystal data are shown in Table 1. Copies of the data can be obtained free of charge at http://www.ccdc.cam.ac.uk/products/csd/request.

Table 1. Selected crystallographic data for all compounds.

Compound	Mg-PABPA	Ca-PABPA	Sr-PABPA	Ni-EDPA	Ni-bpy-EDPA	Mg-TDTMP
Space group	Pc	C2/c	C2/c	P-1	C2/c	P-1
Chemical formula	$C_{14}H_{42}MgN_2O_{18}P_2$	$C_{14}H_{34}CaN_2O_{14}P_2$	$C_{14}H_{34}N_2O_{14}P_2Sr$	$C_2H_{22}NiO_{14}P_2$	$C_{12}H_{26}N_2NiO_{12}P_2$	$C_8H_{34}MgN_2O_{18}P_4$
Formula Mass (g/mol)	612.74	556.45	603.99	390.84	511.00	594.56
λ (Å)	0.71073	0.71073	1.54178	1.54178	0.71073	0.71073
a (Å)	7.2039(17)	29.350(2)	29.511(7)	6.4856(4)	16.953(1)	5.8972(1)
b (Å)	5.9781(13)	6.2111(4)	6.2928(13)	6.5560(3)	14.810(1)	8.9705(1)
c (Å)	31.547(7)	12.9940(7)	13.122(3)	10.1363(5)	10.5126(9)	11.8486(1)
α (°)	90	90	90	89.192(3)	90.000	73.577(1)
β (°)	91.521(9)	107.220(3)	106.947(7)	73.902(3)	127.430(3)	76.201(1)
γ (°)	90	90	90	62.131(3)	90.000	75.534(1)
V (Å^3)	1358.1(5)	2262.5(3)	2331.0(9)	362.66(3)	2096.0(3)	572.58(1)
Crystal size (mm)	0.13 × 0.10 × 0.08	0.12 × 0.10 × 0.10	0.10 × 0.10 × 0.08	0.12 × 0.11 × 0.11	0.37 × 0.13 × 0.12	0.45 × 0.20 × 0.15
Z	2	4	4	1	4	1
ρ_{calc} (g·cm^{-3})	1.498	1.634	1.721	1.790	1.619	1.724
2θ range (°)	2.583–25.043	2.906–27.523	3.131–66.167	4.583–66.836	2.04–38.60	1.82–33.17
Data/Restrains/Parameters	4709/2/339	2604/0/151	2018/0/155	1259/0/99	5872/5/150	4349/8/175
N° reflections	21287	23620	11945	4686	46140	49091
Independent reflections [I > 2σ(I)]	4709	2604	2018	1259	5090	4043
GoF	1.096	1.019	1.137	1.056	1.042	1.443
R Factor [I > 2σ(I)]	[a]R1 = 0.0762, [a]wR2 = 0.1948	[a]R1 = 0.0397, [a]wR2 = 0.0811	[a]R1 = 0.0460, [a]wR2 = 0.1304	[a]R1 = 0.0464, [a]wR2 = 0.1189	[a]R1 = 0.0262, [a]wR2 = 0.0643	[a]R1 = 0.0277, [a]wR2 = 0.0905
R Factor (all data)	[a]R1 = 0.1015, [a]wR2 = 0.2059	[a]R1 = 0.0695, [a]wR2 = 0.0864	[a]R1 = 0.0483, [a]wR2 = 0.1330	[a]R1 = 0.0480, [a]wR2 = 0.1203	[a]R1 = 0.0333, [a]wR2 = 0.0679	[a]R1 = 0.0299, [a]wR2 = 0.0923
CCDC Code	1914866	1914867	1914868	1914870	1914871	1914872

[a] $R_1(F) = \Sigma ||F_o| - |F_c||/\Sigma |F_o|$; $wR_2(F^2) = [\Sigma w(F_o^2 - F_c^2)^2/\Sigma F^4]^{1/2}$.

The structures of **Ni-bpy-EDPA** and **Mg-TDTMP** were solved by employing SHELXTL direct methods and refined by full-matrix least squares on F^2, using the APEX2 software package [31]. All non-H atoms were refined with independent anisotropic displacement parameters. Hydrogen atoms were placed at calculated positions and refined using a riding model, except for the water and phosphonic acid O–H hydrogens, which were located from the Fourier difference density maps and refined using a riding model with O–H distance restraints. Crystallographic details are summarized in Table 1.

2.14. Computational Studies

Electronic Structure Calculations. DFT calculations [32] were performed at the B3LYP/6-31+G* [33–35] level to obtain the distribution of the electronic density ρ of the ligands colored with the Molecular Electrostatic Potential value. Partial charges for all phosphonic oxygen atoms were calculated using the Mulliken [36] and Lowdin [37] Population Analysis. Comparison between the ligands that are used in this study and the non-electron withdrawing group (EWG) analogues should show the deficiency of electron density around the phosphonate groups as a result of the co-existence of the EWGs on each molecule. Furthermore, to better mimic the experimental conditions, the Polarization Continuum Model (PCM) [38] implied in GAMESS(US) [39,40] was used for all the calculations by adding solvent effects. The electronic densities were rendered using wxMacMolPlt [41]. The non EWGs analogs used were ethylphosphonic acid (**H₂EPA**) for **H₄EDPA** and benzylphosphonic acid (**H₂BPA**) for the **H₂PABPA**.

Computational Methodology. The starting geometries for all compounds except **HPABPA^{1-}** were created using the molecular editor Avogadro [42] (Version 1.2.0) and the Universal Force field (UFF) [43]. The next step is the geometry refinement and it was done on the B3LYP/6-31 + G*/PCM level using GAMESS(US) [44–46]. The resulting energetically minimized structures were used for the electronic structure calculations. Since the ligand **HPABPA^{1-}** is found as a dimer in all crystal structures, we used both the monomer and the dimer of **HPABPA^{1-}** for all calculations. Therefore, for the starting geometry of the dimer anion of the ligand, the (**HPABPA^{1-}**)₂ supramolecule was isolated from the CIF file of the **Sr-PABPA** compound using the Avogadro software. This geometry was refined at the B3LYP/6-31 + G*/PCM level using GAMESS(US). The resulting, energetically minimized structure was used for the electronic structure calculations. The geometry of the monomer of **HPABPA^{1-}** was created using Avogadro and the MMFF94 Molecular Force Field created by Merck [47]. The Geometry refinement and the electronic calculations were done at the same level of theory as for the rest compounds.

3. Results

3.1. Synthetic Considerations

One of the major factors affecting metal phosphonate syntheses is pH [48]. In most cases, low pH favors protonation of the phosphonate groups and drastically reduced affinity for metal ions. In contrast, high pH values cause deprotonation, high negative charge on the phosphonate and rapid formation and precipitation of an amorphous metal phosphonate product. Hence, for each metal/phosphonate system there is an optimum pH regime for crystalline product formation [49].

However, even if solution pH ensures electroneutrality, i.e., positive charge from the metal cation equals the negative charge from the phosphonate ligand, the ligand can be found to be non-coordinated (albeit deprotonated), acting simply as the counterion. The metal ion is commonly found bound by water molecules in an aqua complex or coordinated by other ligands present in the system. Other electron-withdrawing, or cationic moieties on the phosphonate ligand backbone may withdraw electron density from the anionic phosphonate group, thus making it less nucleophilic.

The **PABPA** was found as the "free" anion in the structures of **Mg-PABPA**, **Ca-PABPA**, and **Sr-PABPA**. The phosphonate moiety was found to be doubly-deprotonated, whereas the amine group was protonated. This renders the entire ligand monoanionic, hence each divalent metal center requires two **HPABPA** ligands for charge balance. All metal centers were coordinated by water molecules (six in the case of Mg, and eight in the case of Ca and Sr). In spite of the fact that the phosphonate moiety was doubly deprotonated, it remained non-coordinated, apparently due to the decrease in its negative charge because of the $-NH_3^+$ group (this aspect is analyzed in the DFT calculations section, *vide infra*). In this case, water as a ligand was stronger than the phosphonate oxygens.

A similar situation was seen for the **H₆TDTMP^{2-}** tetraphosphonate ligand in **Mg-TDTMP**. Here, every phosphonate moiety was singly-deprotonated, whereas the N atom (a tertiary amine) was found to be protonated. This is the common bis-anionic form of this ligand at the pH region 2–4 [50]. Again, it is assumed that the NH$^+$ moieties deplete the phosphonate moieties of anionic charge, thus making them less nucleophilic.

In the **H₂EDPA^{2-}** ligand, and in the salt **Ni-EDPA**, each phosphonate moiety was singly-deprotonated, thus rendering the ligand a "2−" anion, balancing the "2+" charge of the $[Ni(H_2O)_6]^{2+}$ cation. In spite of the absence of any electron-withdrawing moieties in **H₂EDPA^{2-}** it still remained non-coordinated in the **Ni-EDPA** salt. This has been observed before in the salt $[Ni(2,2'-bpy)(H_2O)_4]\cdot[H_2EDPA]$ [51]. Similarly, in the cationic coordination polymer **Ni-bpy-EDPA** the Ni^{2+} center was coordinated by four water molecules at equatorial positions, whereas the axial positions were occupied by the N atom of the bridging 4,4'-bpy ligand, leaving the **H₂EDPA^{2-}**

diphosphonate to play the role of the counterion for the cationic 1D chain. Here, the 4,4′-bpy ligand displayed strong affinity for the Ni^{2+} center, certainly stronger than the phosphonate oxygens.

3.2. Materials Characterization

Purity of all products was confirmed by elemental analyses (CHN) and powder X-ray diffraction (see Figures S1–S5, in the Supplementary Materials). In some cases (e.g., Mg-PABPA and Ni-EDPA) additional peaks were identified in the XRD diagrams, indicating the presence of impurities, but all our synthetic efforts did not improve product purity.

ATR-IR spectra for all ligands and metal-containing compounds are given in Figure S6, in the Supplementary Materials). For phosphonate-containing materials the region 900–1100 cm^{-1} is commonly used as a "fingerprint" region, in order to confirm changes in the environment of the phosphonate group. Phosphonate bands appeared in this region due to a combination of bands originating from P=O, P-OH and O-P-O vibrations, similar to other reported compounds [52].

For example, for the "free" ligand **H₂PABPA** bands assigned to the –PO_3 moiety appeared at 933, 1064, and 1085 cm^{-1}. In the **Mg-PABPA** compound bands appeared in the same region, but at different positions, namely at 968, 1004, 1058, and 1091 cm^{-1}. The spectra of **Ca-PABPA** and **Sr-PABPA** compounds (essentially identical, since these compounds are isostructural) showed the bands assigned to the –PO_3 moiety in the same region, as expected, but slightly shifted, at 953, 1028, 1061, and 1097 cm^{-1}. In the spectra of the "free" ligand **H₂PABPA** and **Metal-PABPA** compound, the bands due to the aromatic ring were essentially identical. Similar observations can be made for the "Nickel-EDPA" system. The "free" ligand **H₄EDPA** showed bands for the –PO_3 moiety at 921, 953, 993, and 1014 cm^{-1}. Spectra for the **Ni-EDPA** compound showed these bands slightly shifted, at 926, 996, 1029, and 1052 cm^{-1}. Similarly, the **Ni-bpy-EDPA** compound showed very similar bands to those for **Ni-EDPA**, in addition to the strong peaks assigned to the 4,4′-bpy ligand.

Thermogravimetric analyses (TGA) revealed the thermal behavior of the compounds. For example, **Mg-PABPA** (Figure S7 in Supplementary Materials) demonstrated a number of consecutive losses. The first one (~16.5%), starts almost at RT and is completed at ~100 °C. It corresponds to removal of the lattice waters (six calculated, 5.6 found). The second loss (~16.0), starting thereafter was complete at ~150 °C and corresponded to removal of the Mg-coordinated waters (six calculated, 5.5 found). There were some additional small losses at temperatures above 200 °C, which might correspond to the loss of ammonia from the ligand. Similar observations can be noted for **Ca-PABPA** (Figure S8 in Supplementary Materials). The difference here is that **Ca-PABPA** had no lattice water molecules. Hence, the first substantial loss (~23%) completed at ~100 °C corresponded to loss of the Ca-bound waters, followed by some small additional losses that were tentatively assigned to loss of ammonia from the ligand. **Sr-PABPA** (Figure S9 in Supplementary Materials) behaves similarly. Compound **Ni-EDPA** (Figure S10 in Supplementary Materials) loses all its water molecules (six that are Ni-coordinated and one in the lattice) in one single step (~37%) at ~200 °C. Similarly, **Ni-bpy-EDPA** (Figure S11 in Supplementary Materials) showed a one-step loss (~30.3%) of all of its water molecules (four that are Ni-coordinated and one in the lattice). Loss of water molecules will create coordination sites for the phosphonate to bind, provided there is no ligand decomposition. As these phenomena are rather complex, they will be studied in detail and reported in a future publication.

3.3. Crystallographic Description

The structure of **Mg-PABPA** can be described as a salt with $[Mg(H_2O)_6]^{2+}$ being the cation and two monoanionic **HPABPA⁻** ligands being the anions, see Figure 2a. Two **HPABPA⁻** ligands are required for charge balance. There are numerous hydrogen bonds between Mg-bound water molecules, lattice water molecules, the phosphonate moiety and the protonated amine group. The structure (Figure 2b,c) is composed of layers of $[Mg(H_2O)_6]^{2+}$ cations and layers of **HPABPA⁻** ligands that run along the b axis. The orientation of the ligand was *syn-anti*, with the phosphonate groups sitting close to the –NH_3^+ moieties. Lattice water molecules are situated in the vicinity of the Mg-OH_2, -PO_3^{2-} and

−NH_3^+ moieties forming a multitude of hydrogen bonds. The Mg^{2+} center 1s situated in a near perfect octahedral environment, coordinated by six water molecules.

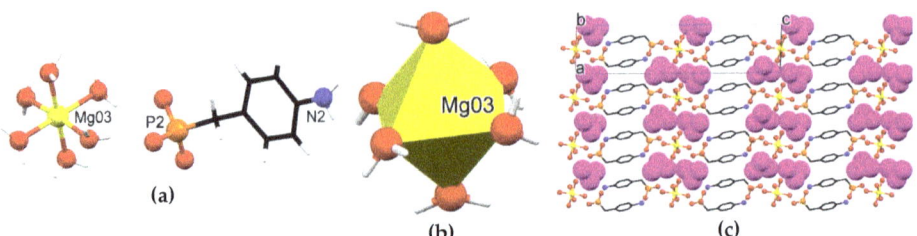

Figure 2. Crystal structure of Mg-PABPA: (**a**) View of the cation-anion pair; (**b**) The $[Mg(H_2O)_6]^{2+}$ polyhedron; (**c**) Packing along the b axis. Lattice waters are shown as magenta spheres.

The salts **Ca-PABPA** and **Sr-PABPA** are isostructural, so only the former is discussed. The structure of **Ca-PABPA** can be described as a salt of $[Ca(H_2O)_8]^{2+}$ aqua complex cation and two monoanionic **HPABPA**$^-$ anions, see Figure 3a. The Ca^{2+} center is coordinated by eight water molecules and is found in a coordination environment best described as bicapped prism, see Figure 3b. In contrast to the **Mg-PABPA** structure, there are no lattice water molecules in the lattice. The structure (Figure 3c) can be described as being composed of layers of $[Ca(H_2O)_8]^{2+}$ cations and layers of **HPABPA**$^-$ ligands that run along the c axis. The orientation of the ligand is *syn-anti*, with the phosphonate groups sitting close to the −NH_3^+ moieties. The -PO_3^{2-} and −NH_3^+ moieties of the **HPABPA**$^-$ ligand form a multitude of hydrogen bonds with the Ca-coordinated water molecules.

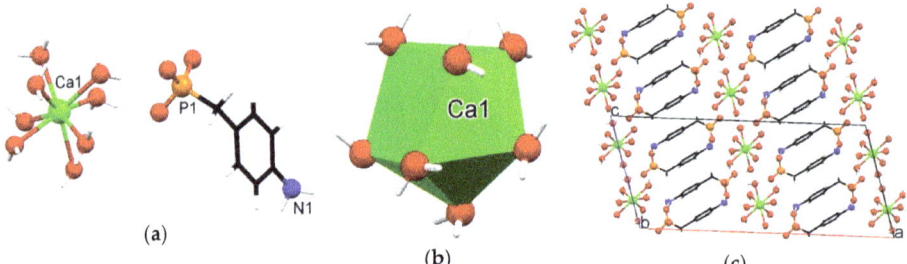

Figure 3. Crystal structure of Ca-PABPA: (**a**) View of the cation-anion pair; (**b**) The $[Ca(H_2O)_8]^{2+}$ polyhedron; (**c**) Packing along the b axis.

The structure of **Mg-TDTMP** can be described as a salt with $[Mg(H_2O)_6]^{2+}$ being the cation and the bis-dianionic H_6TDTMP^{2-} tetraphosphonate ligand being the anion, see Figure 4a. One H_6TDTMP^{2-} ligand is required for charge balance. There are numerous hydrogen bonds between Mg-bound water molecules, the phosphonate moiety, and the tertiary protonated NH^+ group. The structure (Figure 4c) is composed of layers of $[Mg(H_2O)_6]^{2+}$ cations and layers of H_6TDTMP^{2-} ligand that run along the b axis. The orientation of the tetramethylene chain on the ligand is almost parallel to the c axis. There are no lattice water molecules in the structure. The Mg^{2+} center is situated in a near perfect octahedral environment, coordinated by six water molecules (Figure 4b). The salts **Mg-TDTMP** and **Zn-TDTMP** are isostructural [18].

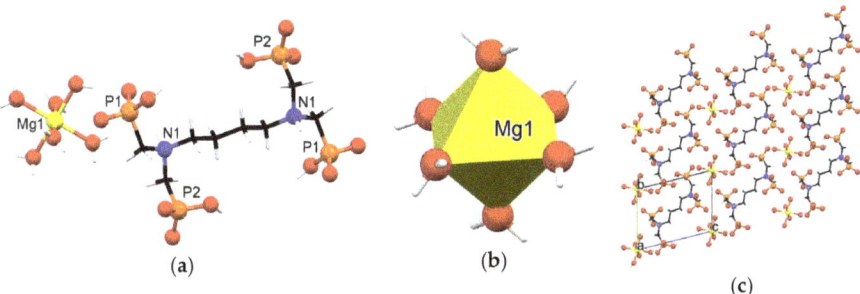

Figure 4. Crystal structure of **Mg-DTTMP**: (a) View of the cation-anion pair; (b) The [Mg(H$_2$O)$_6$]$^{2+}$ polyhedron; (c) Packing along the a axis.

The crystal structure of **Ni-EDPA** is composed of the [Ni(H$_2$O)$_6$]$^{2+}$ dication, the **H$_2$EDPA^{2-}** dianion, and one lattice water molecule, see Figure 5a. The Ni^{2+} center adopts an octahedral geometry, coordinated by six water molecules, see Figure 5b. One **H$_2$EDPA^{2-}** ligand is required for charge balance. There are numerous hydrogen bonds between the Ni-bound water molecules, the lattice water molecule, and the phosphonate moieties. The structure (Figure 5c) can be described as being composed of layers of [Ni(H$_2$O)$_6$]$^{2+}$ cations and layers of the **H$_2$EDPA^{2-}** ligand that run along the b axis. The lattice water molecule is situated in the vicinity of the Ni-OH$_2$, and -PO$_3$H$^-$ moieties, forming a multitude of hydrogen bonds.

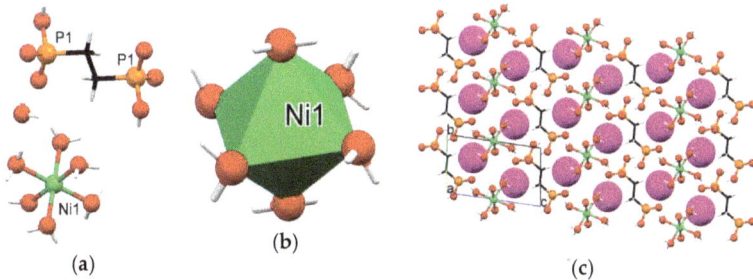

Figure 5. Crystal structure of **Ni-EDPA**: (a) View of the cation-anion pair; (b) The [Ni(H$_2$O)$_6$]$^{2+}$ polyhedron; (c) Packing along the a axis. Lattice waters are shown as magenta spheres.

The structure of **Ni-bpy-EDPA** is different from the ones of other salts described thus far, in that the cationic portion of the salt is a cationic coordination polymer, composed of a cationic *trans*-[Ni(H$_2$O)$_4$(4,4'-bpy)]$^{2+}$ unit, propagating along the *ab* plane (Figure 6a,c). The Ni^{2+} center was coordinated by four water molecules at equatorial positions, whereas the axial positions are occupied by the N atom of the bridging 4,4'-bpy ligand (Figure 6b), leaving the dianionic **H$_2$EDPA^{2-}** diphosphonate to play the role of the counterion for the cationic 1D chain. There is one lattice water molecule (Figure 6c). The **H$_2$EDPA^{2-}** ligand moiety runs parallel to the c axis. There is a complex network of hydrogen bonds that involve the Ni-coordinated waters, the lattice water molecules, and the –PO$_3$H$^-$ moieties. The structure of **Ni-bpy-EDPA** is reminiscent of that in the 1D nickel coordination polymer based on P,P'-diphenylethylenediphosphinic acid and 4,4'-bipyridine [21].

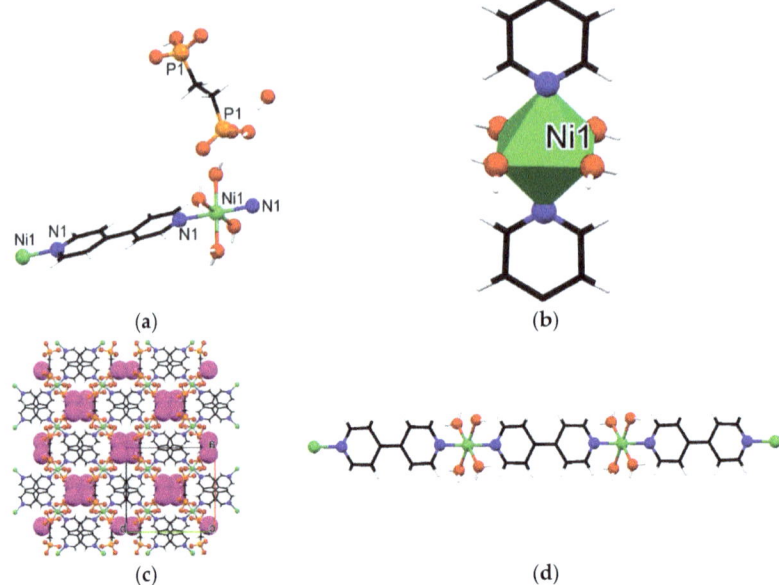

Figure 6. Crystal structure of **Ni-4,4′-bpy-EDPA**: (**a**) View of the cation-anion pair; (**b**) The [Ni(H$_2$O)$_4$(4,4′-bpy)]$^{2+}$ polyhedron; (**c**) Packing along the c axis. Lattice waters are shown as magenta spheres; (**d**) Portion of the 1D "Ni(H$_2$O)$_4$-4,4′-bpy" chain.

3.4. Computational Results: Total Electronic Densities and Partial Charges

In order to visualise the electrostatic potential distribution around the molecular entities, the total electron densities were coloured with the Molecular Electrostatic Potential (MEP) value using the wxMacMolPlt software. All MEP surfaces were constructed using 80 Grid Points and a 0.1 max iso-value for the maps. The RGB surface colouration was used and the colour maps were inverted. The transparency was set to 50 and the surfaces were rendered as smooth and solid. After the energy minimisation of the molecular geometries the partial charges were calculated with Mulliken and Lowdin methods. For this study the partial charges on the phosphonic oxygen atoms are presented, since the oxygen atoms were considered as the coordination centers of the ligands. The mean values of the partial charges per oxygen atom were calculated for each method, and the % reduction of the partial charge was found using Equation (1):

$$\%[Charge\ Reduction] = \frac{<\delta q_{Non-EWG}> - <\delta q_{EWG}>}{<\delta q_{Non-EWG}>} \times 100\% \tag{1}$$

The MEP surfaces and the partial charges of all studied phosphonate molecules are presented below in Table 2 and Figure 7.

From the results above, one can see that comparison of **BPA^{2-}** with either **HPABPA^{1-}** or **(HPABPA^{1-})$_2$** demonstrates the partial charge reduction for both Mulliken and Lowdin methods. The results also show that the formation of the dimer **(HPABPA^{1-})$_2$** from the monomer **HPABPA^{1-}** leads to even greater partial charge reduction, so the formation of the dimer results in lower electron density around the oxygen atoms and in lower probability for coordination to the metal ion. On the other hand, comparison of **HEPA^{1-}** with **H$_2$EDPA^{2-}** demonstrates partial charge reduction for both Mulliken and Lowdin methods. Finally, by comparing **EPA^{2-}** with **H$_2$EDPA^{2-}** a greater partial charge reduction is demonstrated by both methods and in lower probability for coordination to the metal ion.

Table 2. B3LYP/PCM calculated partial charges on the oxygen atoms.

HBPA^{1-}			BPA^{2-}		
Atom #	δq Mulliken	δq Lowdin	Atom #	δq Mulliken	δq Lowdin
16	−0.932911	−0.884627	16	−1.062429	−0.999545
17	−0.969105	−0.891882	17	−1.064163	−1.005116
18	−0.895888	−0.796636	18	−1.072408	−0.986760
mean	−0.932635	−0.857715	mean	−1.066333	−0.997140

(HPABPA^{1-})$_2$			HPABPA^{1-}		
Atom #	δq Mulliken	δq Lowdin	Atom #	δq Mulliken	δq Lowdin
2	−1.011181	−0.817501	15	−1.024967	−0.976540
3	−0.946950	−0.905140	16	−1.062499	−0.994478
4	−0.982415	−0.894151	17	−1.050593	−0.974468
23	−0.975002	−0.898125			
24	−0.936067	−0.911068			
25	−0.974519	−0.821586			
mean	−0.971022	−0.874595	mean	−1.046020	−0.981829

H$_2$EDPA^{2-}		
Atom #	δq Mulliken	δq Lowdin
9	−0.972095	−0.893946
10	−0.961244	−0.900770
11	−0.936159	−0.814461
13	−0.974152	−0.889730
14	−0.959120	−0.811834
16	−0.956155	−0.905960
mean	−0.959821	−0.869450

HEPA^{1-}			EPA^{2-}		
Atom #	δq Mulliken	δq Lowdin	Atom #	δq Mulliken	δq Lowdin
9	−0.965273	−0.893808	9	−1.101121	−1.004788
10	−0.964247	−0.819322	10	−1.103426	−1.014740
12	−0.991020	−0.907261	11	−1.116413	−1.012214
mean	−0.973513	−0.873464	mean	−1.106987	−1.010581

Molecules Pair	% Charge Reduction	
	Mulliken	Lowdin
HBPA^{1-} - HPABPA^{1-}	−12%	−15%
BPA^{2-} - HPABPA^{1-}	+2%	+2%
HBPA^{1-} - (HPABPA^{1-})$_2$	−4%	−2%
BPA^{2-} - (HPABPA^{1-})$_2$	+9%	+12%
HPABPA^{1-} - (HPABPA^{1-})$_2$	+7%	+11%
HEPA^{1-} - H$_2$EDPA^{2-}	+1%	+1%
EPA^{2-} - H$_2$EDPA^{2-}	+13%	+14%

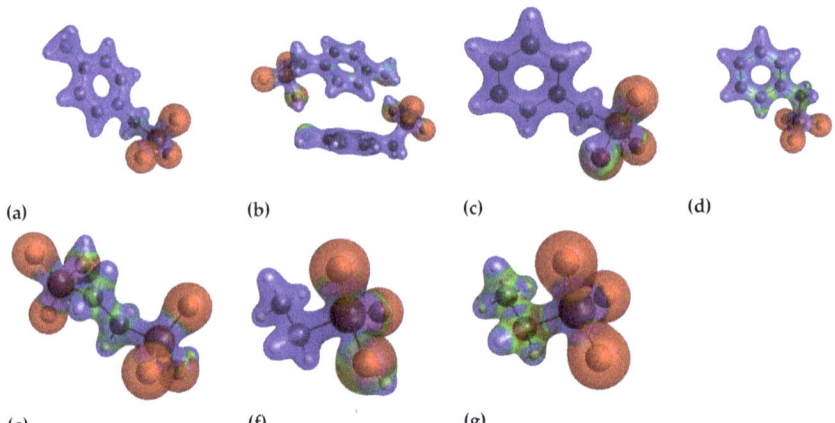

Figure 7. DFT calculated total electron density colored with the Electrostatic Potential Map value for a) **HBPA^{1-}**, b) **(HPABPA^{1-})$_2$**, c) **HPABPA^{1-}** d) **BPA^{2-}** e) **H$_2$EDPA^{2-}**, f) **HEPA^{1-}**, and g) **EPA^{2-}**, at the B3LYP/PCM geometry.

4. Discussion

It is common in metal phosphonate chemistry to seek for MOFs, coordination polymers, and networks that represent new and exciting structures in which the metal-phosphonate bonds play a key role. However, there is a class of metal phosphonate compounds (salts) where the deprotonated (anionic) phosphonate plays the role of the counterion to a cationic metal complex, or to a cationic coordination polymer. Such non-coordinating systems are useful in modeling phosphonate compounds in biological [53], polymeric, [54], or hydrogel matrices [6,55].

In the ionic metal phosphonates presented herein, there are two major structural features that draw attention: (a) the cationic metal center and its coordination environment, and (b) the deprotonated anionic phosphonate. The metal cation is bound by aqua ligands (solvent), and occasionally, by other, externally-added ligands (4,4′-bpy in this case for **Ni-bpy-EDPA**). As reported before, there is a correlation between the ionic radius of the metal ion and the bond distances between the metal and the ligand atoms coordinated to the metal center [2,56–58]. Since there is no direct metal-phosphonate bonding in the compounds reported herein, the only correlations that can be drawn are those for metal-O$_{water}$ bonds. By calculating the average metal-O$_{water}$ bond lengths, an increasing trend can be observed, see Figure 8.

Metal-O$_{water}$ bond lengths for the divalent metal ions Zn [18], Ba [58], and Co [12] were taken from literature sources, whereas those for Mg, Ni, Ca and Sr were taken from the structures described in the present paper.

The phosphonate anion is invariably stabilized by hydrogen bonds. For the "simple" system **Ni-EDPA** (i.e., the phosphonate ligand contains only phosphonate groups, without other ligands present) the H$_2$EDPA^{2-} dianion interacts with a total of 16 hydrogen bonds (with neighboring H$_2$EDPA^{2-} dianions and water molecules). Similar observations can be made for the system **Ni-bpy-EDPA**. The HPABPA$^-$ ligand possesses a doubly-deprotonated phosphonate group, but a protonated amine group. The –PO$_3$$^{2-}$ moiety in the isostructural compounds **Ca-PABPA** and **Sr-PABPA** participates in seven hydrogen bonds, whereas the –NH$_3$$^+$ moiety participates in three. In the structure of **Mg-PABPA**, where there are also lattice waters, the –PO$_3$$^{2-}$ moiety participates in five hydrogen bonds, whereas the –NH$_3$$^+$ moiety participates in three. Here, the lattice and Mg-coordinated waters form a multitude of hydrogen bonding interactions. Finally, in the **Mg-TDTMP** salt, the four phosphonate groups, overall, participate in twenty one hydrogen bonds, and each protonated tertiary amine group, in one. In general, it has been observed that the higher the coordination state of the phosphonate ligand, the lower is the extent of hydrogen bonding [2].

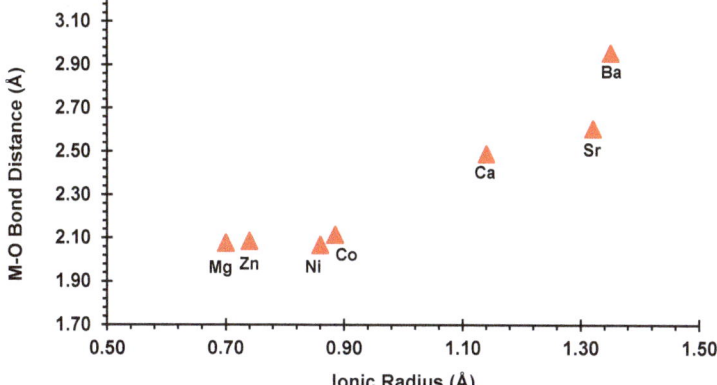

Figure 8. Dependence of M-O(H$_2$O) bond lengths in divalent metal-aqua species on metal ionic radius.

5. Conclusions

Herein, we presented a family of metal phosphonate-based inorganic-organic salts, which share a common characteristic, i.e., the phosphonate anion is not coordinated to a metal center, but merely plays the role of the counter-anion. In contrast to the majority of metal phosphonate materials reported in the literature, in which the phosphonate ligand was found coordinated to the metal center, the compounds **Mg-PABPA, Ca-PABPA, Sr-PABPA, Ni-EDPA, Mg-TDTMP**, and **Ni-bpy-EDPA** contained the metal center as a hydrated complex, except **Ni-bpy-EDPA**, which contained the Ni^{2+} center as part of a coordination polymer with the 4,4'-bpy bridging ligand. The formation of these "platonic" compounds can be explained based on the high affinity of the M^{2+} metal ions for water as a ligand (or for 4,4'-bpy) and the reduced affinity of the phosphonate ligand, particularly in the case of **H$_2$PABPA**, for the metal centers. Another factor that plays a role in the stabilization of such compounds in the solid state is the multitude of hydrogen bonding interactions that occur, due to the plethora of hydrogen bonding donors and acceptors. Hence, these compounds could be attractive for applications such as proton conductivity, and such studies are currently under way in our laboratories.

Although the main focus in metal phosphonate chemistry is the synthesis and exploration of coordination polymers [59], the class of ionic phosphonates (either with metal complexes, or organic counter-cations) could potentially unlock new potential in this research area. Given the plethora of structurally diverse (poly)phosphonate [60,61] and "mixed" phosphonate (e.g. with carboxy- [56], sulfono- [62], amino- [63], or N-heterocyclic groups [64]) ligands available, one could envision countless possibilities in the synthesis, fabrication and properties of such solids [65].

Supplementary Materials: The following are available online at http://www.mdpi.com/2073-4352/9/6/301/s1, Figures S1–S5: Calculated and measured powder XRD diagrams for all compounds; Figure S6: ATR-IR spectra of all metal phosphonate compounds and "free" ligands; Figures S7–S11: TGA and DSC traces for all compounds; Crystallographic Information Files (six cif files) for all compounds.

Author Contributions: conceptualization, K.D.D.; software, K.X.; validation, K.X.; investigation, Z.A., S.C., D.C.-L., G.M., J.Z, and J.Z.; writing—original draft preparation, K.D.D.; writing—review and editing, all authors; supervision, K.D.D.

Funding: K.D.D. acknowledges the Research Committee of the University of Crete (Grant KA 10329) for financial support.

Acknowledgments: We thank Prof. Aurelio Cabeza (Universidad de Malaga, Spain) for technical support.

Conflicts of Interest: The authors declare no conflict of interest.

References

1. Clearfield, A.; Demadis, K.D. *Metal Phosphonate Chemistry: From Synthesis to Applications*; Royal Society of Chemistry: London, UK, 2012.
2. Demadis, K.D.; Stavgianoudaki, N. Structural diversity in metal phosphonate frameworks: Impact on applications. In *Metal Phosphonate Chemistry: From Synthesis to Applications*; Clearfield, A., Demadis, K.D., Eds.; Royal Society of Chemistry: London, UK, 2012; Chapter 14; pp. 438–492.
3. Zaręba, J.K. Tetraphenylmethane and tetraphenylsilane as building units of coordination polymers and supramolecular networks–A focus on tetraphosphonates. *Inorg. Chem. Commun.* **2017**, *86*, 172–186. [CrossRef]
4. Bao, S.-S.; Shimizu, G.K.H.; Zheng, L.-M. Proton conductive metal phosphonate frameworks. *Coord. Chem. Rev.* **2019**, *378*, 577–594. [CrossRef]
5. Groves, J.A.; Miller, S.R.; Warrender, S.J.; Mellot-Draznieks, C.; Lightfoot, P.; Wright, P.A. The first route to large pore metal phosphonates. *Chem. Commun.* **2006**, *31*, 3305–3307. [CrossRef] [PubMed]
6. Papathanasiou, K.E.; Vassaki, M.; Spinthaki, A.; Alatzoglou, F.-E.G.; Tripodianos, E.; Turhanen, P.; Demadis, K.D. Phosphorus chemistry: From small molecules to polymers to pharmaceutical and industrial applications. *Pure Appl. Chem.* **2019**, *91*, 421–441. [CrossRef]
7. Shah, B.; Chudasama, U. Application of zirconium phosphonate—a novel hybrid material as an ion exchanger. *Des. Wat. Treat.* **2012**, *38*, 227–235.
8. Armakola, E.; Colodrero, R.M.P.; Bazaga-García, M.; Salcedo, I.R.; Choquesillo-Lazarte, D.; Cabeza, A.; Kirillova, M.V.; Kirillov, A.M.; Demadis, K.D. Three-component copper-phosphonate-auxiliary ligand systems: Proton conductors and efficient catalysts in mild oxidative functionalization of cycloalkanes. *Inorg. Chem.* **2018**, *57*, 10656–10666. [CrossRef] [PubMed]
9. Moschona, A.; Plesu, N.; Mezei, G.; Thomas, A.; Demadis, K.D. Corrosion protection of carbon steel by tetraphosphonates of systematically different molecular size. *Corr. Sci.* **2018**, *145*, 135–150. [CrossRef]
10. Gałęzowska, J.; Kafarski, P.; Kozłowski, H.; Młynarz, P.; Nurchi, V.M.; Pivetta, T. N,N'-ethylenediaminobis(benzylphosphonic acids) as a potent class of chelators for metal ions. *Inorg. Chim. Acta* **2009**, *362*, 707–713. [CrossRef]
11. Heering, C.; Nateghi, B.; Janiak, C. Charge-assisted hydrogen-bonded networks of NH_4^+ and $[Co(NH_3)_6]^{3+}$ with the new linker anion of 4-phosphono-biphenyl-4'-carboxylic acid. *Crystals* **2016**, *6*, 22. [CrossRef]
12. Wilk, M.; Janczak, J.; Videnova-Adrabinska, V. Hexaaquacobalt(II) bis[hydrogen bis(4-carboxyphenylphosphonate)] dihydrate. *Acta Crystallogr. Sect. C—Cryst. Struct. Commun.* **2011**, *67*, 9–12. [CrossRef]
13. Gholivand, K.; Farrokhi, A.R. Supramolecular hydrogen-bonded frameworks from a new bisphosphonic Acid and transition metal ions. *Z. Anorg. Allg. Chem.* **2011**, *637*, 263–268. [CrossRef]
14. Sergienko, V.S.; Aleksandrov, G.G.; Afonin, E.G. An unusual function of the anion of 1-hydroxyethane-1,1-diphosphonic acid (H_4L): Crystal structure of $[Ni(phen)_3](H_7L_2)_{0.5}(H_5L_2)_{0.5}\cdot 2H_2O$. *Cryst. Rep.* **2000**, *45*, 432–438. [CrossRef]
15. Yang, J.; Ma, J.-F.; Zheng, G.-L.; Li, L.; Li, F.-F.; Zhang, Y.-M.; Liu, J.-F. Divalent transition metal phosphonates with new structure containing hydrogen-bonded layers of phosphonate anions. *J. Solid State Chem.* **2003**, *174*, 116–124. [CrossRef]
16. Latham, K.; Coyle, A.M.; Rix, C.J.; Fowless, A.; White, J.M. Effect of ring substituents on crystal packing and H-bonding in a series of halobis(phen)copper(II) arylphosphonic acid complexes. *Polyhedron* **2007**, *26*, 222–236. [CrossRef]
17. Ma, K.-R.; Wei, C.-L.; Zhang, Y.; Kan, Y.-H.; Cong, M.-H.; Yang, X.-J. Structures and spectroscopy studies of two M(II)-phosphonate coordination polymers based on alkaline earth metals (M = Ba, Mg). *J. Spectroscopy* **2013**, *2013*. [CrossRef]
18. Demadis, K.D.; Barouda, E.; Zhao, H.; Raptis, R.G. Structural architectures of charge-assisted, hydrogen-bonded, 2D layered amine⋯tetraphosphonate and zinc⋯tetraphosphonate ionic materials. *Polyhedron* **2009**, *28*, 3361–3367. [CrossRef]
19. Murugavel, R.; Kuppuswamy, S.; Randoll, S. Cooperative binding of phosphate anion and a neutral nitrogen donor to alkaline-earth metal ions. Investigation of Group 2 metal-organophosphate interaction in the absence and presence of 1,10-phenanthroline. *Inorg. Chem.* **2008**, *47*, 6028–6039. [CrossRef]

20. Lee, B.H.; Lynch, V.M.; Cao, G.; Mallouk, T.E. Structure of [Mg{HO$_3$PCH(C$_6$H$_5$)$_2$}$_2$]·8H$_2$O, a layered phosphonate salt. *Acta Cryst.* **1988**, *C44*, 365–367.
21. Guerri, A.; Taddei, M.; Bataille, T.; Moneti, S.; Boulon, M.-E.; Sangregorio, C.; Costantino, F.; Ienco, A. Same not the same: Thermally driven transformation of nickel phosphinate-bipyridine one-dimensional chains into three-dimensional coordination polymers. *Cryst. Growth Des.* **2018**, *18*, 2234–2242. [CrossRef]
22. Schier, A.; Gamper, S.; Müller, G. Synthesis and crystal structure of magnesium bis[2-aminoethyl(hydrogen)phosphonate] octahydrate, Mg(2-AEPH)$_2$·8H$_2$O. *Inorg. Chim. Acta* **1990**, *177*, 179–183. [CrossRef]
23. Smith, G.; Wermuth, U.D. The coordination complex structures and hydrogen bonding in the three-dimensional alkaline earth metal salts (Mg, Ca, Sr and Ba) of (4-aminophenyl)arsonic acid. *Acta Cryst.* **2017**, *C73*, 61–67. [CrossRef] [PubMed]
24. Nguyen, T.V.; Bekensir, A. Aromatic cation activation: Nucleophilic substitution of alcohols and carboxylic acids. *Org. Lett.* **2014**, *16*, 1720–1723. [CrossRef] [PubMed]
25. Taylor, S.D.; Kotoris, C.C.; Dinaut, A.N.; Chen, M.-J. Synthesis of aryl(difluoromethylenephosphonates) via electrophilic fluorination of α-carbanions of benzylic phosphonates with N-fluorobenzenesulfonimide. *Tetrahedron* **1998**, *54*, 1691–1714. [CrossRef]
26. Penicaud, V.; Maillet, C.; Janvier, P.; Pipelier, M.; Bujoli, B. New water-soluble diamine complexes as catalysts for the hydrogenation of ketones under hydrogen pressure. *Eur. J. Org. Chem.* **1999**, *7*, 1745–1748.
27. Wydysh, E.A.; Medghalchi, S.M.; Vadlamudi, A.; Townsendd, C.A. Design and synthesis of small molecule glycerol 3-phosphate acyltransferase inhibitors. *J. Med. Chem.* **2009**, *52*, 3317–3327. [CrossRef] [PubMed]
28. *APEX3, V2018.7-2*; Bruker AXS, Inc.: Madison, WI, USA, 2018.
29. Sheldrick, G.M. A short history of SHELX. *Acta Cryst.* **2008**, *A64*, 339–341.
30. Sheldrick, G.M. Crystal structure refinement with SHELXL. *Acta Cryst.* **2015**, *C71*, 3–8.
31. *APEX2 v2014.9-0*; Bruker AXS Inc.: Madison, WI, USA, 2014.
32. Kohn, W.; Sham, L.J. Self-consistent equations including exchange and correlation effects. *Phys. Rev.* **1965**, *140*, A1133–A1138. [CrossRef]
33. Becke, A.D. Density-functional thermochemistry. III. The role of exact exchange. *J. Chem. Phys.* **1993**, *98*, 5648–5653. [CrossRef]
34. King, H.F.; Dupuis, M. Numerical integration using rys polynomials. *J. Comput. Phys.* **1976**, *21*, 144–165. [CrossRef]
35. Hariharan, P.C.; Pople, J.A. The influence of polarization functions on molecular orbital hydrogenation energies. *Theoret. Chim. Acta* **1973**, *28*, 213–222. [CrossRef]
36. Mulliken, R.S. Electronic population analysis on LCAO–MO molecular wave functions. Parts 1-4. *J. Chem. Phys.* **1955**, *23*, 1833–1840, 1841–1846, 2338–2342, and 2343–2346. [CrossRef]
37. Lowdin, P.-O. On the orthogonality problem. *Adv. Chem. Phys.* **1970**, *5*, 185–199.
38. Li, H.; Pomelli, C.; Jensen, J.H. Continuum solvation of large molecules described by QM/MM: A semi-iterative implementation of the PCM/EFP interface. *J. Theor. Chem. Acc.* **2003**, *109*, 71–84. [CrossRef]
39. Schmidt, M.W.; Baldridge, K.K.; Boatz, J.A.; Elbert, S.T.; Gordon, M.S.; Jensen, J.H.; Koseki, S.; Matsunaga, N.; Nguyen, K.A.; Su, S.; et al. General atomic and molecular electronic structure system. *J. Comput. Chem.* **1993**, *14*, 1347–1363. [CrossRef]
40. Gordon, M.S.; Schmidt, M.W. Advances in electronic structure theory: GAMESS a decade later. In *Theory and Applications of Computational Chemistry, the First Forty Years*; Dykstra, C.E., Frenking, G., Kim, K.S., Scuseria, G.E., Eds.; Elsevier: Amsterdam, The Netherland, 2005; Chapter 41; pp. 1167–1189.
41. Bode, B.M.; Gordon, M.S. MacMolPlt: a graphical user interface for GAMESS. *J. Mol. Graphics Mod.* **1998**, *16*, 133–138. [CrossRef]
42. Hanwell, M.D.; Curtis, D.E.; Lonie, D.C.; Vandermeersch, T.; Zurek, E.; Hutchison, G.R. Avogadro: an advanced semantic chemical editor, visualization, and analysis platform. *J. Cheminform.* **2012**, *4*, 17. [CrossRef] [PubMed]
43. Rappe, A.K.; Casewit, C.J.; Colwell, K.S.; Goddard III, W.A.; Skiff, W.M. UFF, a full periodic table force field for molecular mechanics and molecular dynamics simulations. *J. Am. Chem. Soc.* **1992**, *114*, 10024–10035. [CrossRef]
44. Baker, J. An algorithm for the location of transition states. *J. Comput. Chem.* **1986**, *7*, 385–395. [CrossRef]

45. Helgaker, T. Transition-state optimizations by trust-region image minimization. *Chem. Phys. Lett.* **1991**, *182*, 503–510. [CrossRef]
46. Culot, P.; Dive, G.; Nguyen, V.H.; Ghuysen, J.M. A quasi-Newton algorithm for first-order saddle-point location. *Theoret. Chim. Acta* **1992**, *82*, 189–205. [CrossRef]
47. Halgren, T.A. Merck molecular force field. I. Basis, form, scope, parameterization, and performance of MMFF94. *J. Comput. Chem.* **1996**, *17*, 490–519. [CrossRef]
48. Demadis, K.D.; Papadaki, M.; Raptis, R.G.; Zhao, H. Corrugated, sheet-like architectures in layered alkaline earth metal R,S-hydroxyphosphonoacetate frameworks: Applications for anti-corrosion protection of metal surfaces. *Chem. Mater.* **2008**, *20*, 4835–4846. [CrossRef]
49. Lodhia, S.; Turner, A.; Papadaki, M.; Demadis, K.D.; Hix, G.B. Polymorphism, composition and structural variability in topology in 1D, 2D and 3D copper phosphonocarboxylate materials. *Cryst. Growth Des.* **2009**, *9*, 1811–1822. [CrossRef]
50. Ruiz-Agudo, E.; Rodriguez-Navarro, C.; Sebastian-Pardo, E. Sodium sulfate crystallization in the presence of phosphonates: Implications in ornamental stone conservation. *Cryst. Growth Des.* **2006**, *6*, 1575–1583. [CrossRef]
51. Demadis, K.D.; Anagnostou, Z.; Panera, A.; Mezei, G.; Kirillova, M.V.; Kirillov, A.M. Three-Component 1D and 2D metal phosphonates: Structural variability, topological analysis and catalytic hydrocarboxylation of alkanes. *RSC-Adv.* **2017**, *7*, 17788–17799. [CrossRef]
52. Demadis, K.D.; Armakola, E.; Papathanasiou, K.E.; Mezei, G.; Kirillov, A.M. Structural systematics and topological analysis of coordination polymers with divalent metals and a glycine-derived tripodal phosphonocarboxylate. *Cryst. Growth Des.* **2014**, *14*, 5234–5243. [CrossRef]
53. Papathanasiou, K.E.; Demadis, K.D. Phosphonates in matrices. In *Tailored Organic-Inorganic Materials*; Brunet, E., Clearfield, A., Colon, J.L., Eds.; John Wiley & Sons Inc.: New York, NY, USA, 2015; Chapter 3; pp. 83–135.
54. Demadis, K.D.; Theodorou, I.; Paspalaki, M. Controlled release of bis-phosphonate pharmaceuticals from cationic biodegradable polymeric matrices. *Ind. Eng. Chem. Res.* **2011**, *50*, 5873–5876. [CrossRef]
55. Papathanasiou, K.E.; Turhanen, P.; Brückner, S.I.; Brunner, E.; Demadis, K.D. Smart, programmable and responsive injectable hydrogels for controlled release of cargo osteoporosis drugs. *Sci. Rep.* **2017**, *7*. [CrossRef]
56. Bazaga-García, M.; Papadaki, M.; Colodrero, R.M.P.; Olivera-Pastor, P.; Losilla, E.R.; Nieto-Ortega, B.; Aranda, M.A.G.; Choquesillo-Lazarte, D.; Cabeza, A.; Demadis, K.D. Tuning proton conductivity in alkali metal phosphonocarboxylates by cation size-induced and water-facilitated proton transfer pathways. *Chem. Mater.* **2015**, *27*, 424–435. [CrossRef]
57. Salcedo, I.R.; Colodrero, R.M.P.; Bazaga-García, M.; Vasileiou, A.; Papadaki, M.; Olivera-Pastor, P.; Infantes-Molina, A.; Losilla, E.R.; Mezei, G.; Cabeza, A.; et al. From light to heavy alkali metal tetraphosphonates (M = Li, Na, K, Rb, Cs): Cation size-induced structural diversity and water-facilitated proton conductivity. *CrystEngComm* **2018**, *20*, 7648–7658. [CrossRef]
58. Demadis, K.D.; Katarachia, S.D.; Zhao, H.; Raptis, R.G.; Baran, P. Alkaline earth metal organotriphosphonates: Inorganic-organic polymeric hybrids from dication-dianion association. *Cryst. Growth Des.* **2006**, *6*, 836–838. [CrossRef]
59. Zorlu, Y.; Erbahar, D.; Çetinkaya, A.; Bulut, A.; Erkal, T.S.; Yazaydin, A.O.; Beckmann, J.; Yücesan, G. A cobalt arylphosphonate MOF – superior stability, sorption and magnetism. *Chem. Commun.* **2019**, *55*, 3053–3056. [CrossRef]
60. Yücesan, G.; Zorlu, Y.; Stricker, M.; Beckmann, J. Metal-organic solids derived from arylphosphonic acids. *Coord. Chem. Rev.* **2018**, *369*, 105–122. [CrossRef]
61. Silvestre, J.-P.; Dao, N.Q.; Leroux, Y. A survey of the behavior of the hydroxybisphosphonic function in crystallized acids, metallic salts, and some related compounds. *Heteroatom Chem.* **2001**, *12*, 73–89. [CrossRef]
62. Beyer, O.; Homburg, T.; Albat, M.; Stock, N.; Lüning, U. Synthesis of phosphonosulfonic acid building blocks as linkers for coordination polymers. *New J. Chem.* **2017**, *41*, 8870–8876. [CrossRef]
63. Schmidt, C.; Feyand, M.; Rothkirch, A.; Stock, N. High-throughput and in situ EDXRD investigation on the formation of two new metal aminoethylphosphonates–Ca($O_3PC_2H_4NH_2$) and Ca(OH)($O_3PC_2H_4NH_3$)·$2H_2O$. *J. Solid State Chem.* **2012**, *188*, 44–49. [CrossRef]

64. Gałęzowska, J.; Czapor-Irzabek, H.; Janicki, R.; Chmielewska, E.; Janek, T. New aspects of coordination chemistry and biological activity of NTMP-related diphosphonates containing a heterocyclic ring. *New J. Chem.* **2017**, *41*, 10731–10741. [CrossRef]
65. Shearan, S.; Stock, N.; Emmerling, F.; Demel, J.; Wright, P.A.; Demadis, K.D.; Vassaki, M.; Costantino, F.; Vivani, R.; Sallard, S.; et al. New directions in metal phosphonate and phosphinate chemistry. *Crystals* **2019**, *9*, 270. [CrossRef]

© 2019 by the authors. Licensee MDPI, Basel, Switzerland. This article is an open access article distributed under the terms and conditions of the Creative Commons Attribution (CC BY) license (http://creativecommons.org/licenses/by/4.0/).

Article

Novel Cerium Bisphosphinate Coordination Polymer and Unconventional Metal–Organic Framework

Jan Rohlíček [1], Daniel Bůžek [2], Petr Brázda [1], Libor Kobera [3], Jan Hynek [2], Jiří Brus [3], Kamil Lang [2] and Jan Demel [2,*]

[1] Institute of Physics of the Czech Academy of Sciences, 18221 Prague, Czech Republic; rohlicek@fzu.cz (J.R.); brazda@iic.cas.cz (P.B.)
[2] Institute of Inorganic Chemistry of the Czech Academy of Sciences, Husinec-Řež 1001, 250 68 Řež, Czech Republic; buzek@iic.cas.cz (D.B.); hynek@iic.cas.cz (J.H.); lang@iic.cas.cz (K.L.)
[3] Institute of Macromolecular Chemistry of the Czech Academy of Sciences, Heyrovského nám. 2, 162 06 Prague 6, Czech Republic; kobera@imc.cas.cz (L.K.); brus@imc.cas.cz (J.B.)
* Correspondence: demel@iic.cas.cz

Received: 30 April 2019; Accepted: 7 June 2019; Published: 12 June 2019

Abstract: The first Ce(III)-based coordination polymer ICR-9 (ICR stands for Inorganic Chemistry Řež), with the formula $Ce_2(C_8H_{10}P_2O_4)_3$, containing ditopic phenylene-1,4-bis(methylphosphinic acid) linker, was synthetized under solvothermal conditions. The crystal structure, solved using electron diffraction tomography (EDT), revealed 2D layers of octahedrally coordinated cerium atoms attached together through O-P-O bridges. The structure is nonporous, however, the modification of synthetic conditions led to unconventional metal–organic framework (or defective amorphous phase) with a specific surface area up to approximately 400 $m^2\ g^{-1}$.

Keywords: coordination polymer; Cerium; defects; amorphous; porosity; electron diffraction tomography; solid state NMR

1. Introduction

Coordination polymers, and particularly the porous subgroup referred to as metal–organic frameworks (MOFs), have been extensively studied during the past decades [1]. Since the early years, carboxylate linkers were in the spotlight, however, the high specific surface area and tunability of the structures is accompanied with generally low stability in air and aqueous environments, hindering their industrial applications [2–4].

The next logical step was to use linkers based on a phosphonate group (RPO_3^{2-}) which forms stronger coordination bonds than carboxylates, however, the presence of three coordinating oxygen atoms results in many coordination modes and the resulting structures are often nonporous [5]. Despite progress in recent years, well crystalline porous structures are difficult to prepare [6], often utilizing high-throughput methods [5]. We kindly refer the reader to other papers included in this issue for more details [7]. On the other hand, syntheses of poorly crystalline or amorphous metal phosphonates, displaying porosity due to the presence of defects, referred to as unconventional metal–organic frameworks (UMOFs), have yielded a large number of highly stable materials [8].

Interestingly, phosphinate-(R^1R^2POOH) based coordination polymers have attracted much less attention [9]. Most of the reported phosphinate-based coordination polymers use (i) monophosphinic acids, e.g., diphenylphosphinic acid [10], ethylbutylphosphinic acid [11], or m-carboranylphosphininc acid [12,13] which act as bridging ligands forming 1D infinite chains; (ii) ligands with both carboxylic and phosphinic groups, e.g., (2-carboxyethyl)(phenyl)phosphinic acid which form 2D or 3D networks [14–16]; (iii) ligands bearing two phosphinic acid groups separated either by methylene bridge forming a 2D network [17], or by 1,1'-substituted ferrocene forming a flexible polymer chain [18,19]. Recently, we reported on the first permanently porous MOF made of bisphosphinate linkers [20].

Cerium in oxidation states Ce(III) and Ce(IV) is known to form stable salts. For this reason, cerium salts in both oxidation states were used for the preparation of coordination polymers, including MOFs. The first Ce-MOF was prepared with phosphonate linker [21], later several studies described the synthesis of cerium analogues of carboxylate Zr-MOFs composed of the $[Ce_6O_4(OH)_4]^{12+}$ secondary building units [22,23]. MOFs based on Ce(III) forming a linear inorganic building units were also reported [24]. For a recent review on Ce-MOFs, please see [25]. Interestingly, phosphinate coordination polymers with cerium have not been reported yet.

Here, we report on the synthesis and crystal structure of ICR-9 (ICR stands for Inorganic Chemistry Řež), a Ce(III) coordination polymer with phenylene-1,4-bis(methylphosphinic acid)–H_2PBP(Me) linker. Because of the microcrystalline nature of the polymer, the crystal structure was determined using electron diffraction tomography (EDT). The pores in the structure are too small to accommodate gas molecules; however, when defective amorphous phase (UMOF) is formed along with the ICR-9 phase, the phase mixtures become microporous with a specific surface area up to approximately 400 $m^2\ g^{-1}$.

2. Materials and Methods

2.1. Materials

Ammonium cerium(IV) nitrate (>99.99%), cerium(III) chloride heptahydrate (99.9%), and cerium(III) nitrate hexahydrate (99%) were purchased from Sigma–Aldrich. Phenylene-1,4-bis(methylphosphinic acid) (H_2PBP(Me)) was synthetized according to a procedure described earlier [20]. N,N-Dimethylformamide (Penta, Czech Republic; abbreviated as DMF) and acetone (Lach-Ner, Czech Republic) were used as received. Reactions were performed using deionized water (conductivity < 0.15 $\mu S\ cm^{-1}$).

2.2. Preparation of ICR-9

2.2.1. Synthesis of Well-Crystalline ICR-9

A Schlenk tube was charged with 40 mg (0.171 mmol) of H_2PBP(Me) and 8 mL of water. The mixture was preheated in an oil bath at 100 °C under stirring. Then, the solution of 93.7 mg (0.171 mmol) of ammonium cerium(IV) nitrate dissolved in 4 mL of DMF was added. The Schlenk tube was sealed and the crystallization was carried out at 100 °C for 3 h under continuous stirring. The resulting white solid was centrifuged (Hettich Rotina 380 R, 5 min, 11,000 rpm), washed three times with DMF and five times with acetone, and air-dried, yielding 38 mg of yellow powder, denoted as ICR-9Cryst. Elemental analysis CHN calculated (%) for $Ce_2(C_8H_{10}P_2O_4)_3$: C 28.70, H 3.10, N 0.0; found C 29.13, H 3.08, N 0.0.

Well-crystalline ICR-9 was also prepared from Ce(III) salts—cerium chloride and nitrate. The procedure was identical to the one described above, using 63.7 mg $CeCl_3 \cdot 7H_2O$ or 74.3 mg $Ce(NO_3)_3 \cdot 6H_2O$ (both 0.171 mmol) instead of ammonium cerium(IV) nitrate.

2.2.2. Synthesis of ICR-9 with UMOF Phase

The synthetic protocol was similar to the one used for the well-crystalline ICR-9, however, the solvents for preparing the reactant solutions were interchanged. Thus, a Schlenk tube was charged with 40 mg (0.171 mmol) of H_2PBP(Me) and 8 mL of DMF. The mixture was preheated in an oil bath at 100 °C under stirring. Then, the solution of 93.7 mg (0.171 mmol) of ammonium cerium(IV) nitrate dissolved in 4 mL of water was added. The Schlenk tube was sealed and the crystallization was carried out at 100 °C for 0.5 and 1.5 h under continuous stirring to form ICR-9A and ICR-9B, respectively. The resulting white solids were centrifuged (Hettich Rotina 380 R, 5 min, 11,000 rpm), washed three times with DMF and five times with acetone, and air-dried. The yield was 35 and 41 mg for ICR-9A and ICR-9B, respectively. In the case of ICR-9A, the BET (Brunauer–Emmett–Teller) specific surface area varied from batch to batch as much as ±10%. On the other hand, the syntheses of ICR-9B were well reproducible with batch to batch BET specific surface differences within the experimental error. Elemental analysis CHN calculated (%) for $Ce_2(C_8H_{10}P_2O_4)_3$: C 28.70, H 3.10, N 0.0; ICR-9A found C 24.83, H 3.15, N 1.29; ICR-9B found C 25.22, H 3.00, N 0.74. The content of nitrogen is in line with a small amount of DMF trapped in the pores of UMOF, see below. The lower content of measured carbon can originate from defects in the structure of the UMOF phase.

2.3. Instrumental Methods

Electron diffraction tomography experiments were performed on a Philips CM120 microscope (120 kV) with a LaB_6 cathode equipped with an Olympus SIS Veleta CCD camera (14 bit) (Olympus Corporation, Tokyo, Japan). Samples were measured at 100 K (sample holder tip temperature). The platelet crystals were twisted and preferentially oriented with [001] parallel to the electron beam. Due to large mosaicity and a lattice parameter c of 40.7 Å, it was not possible to use precession [26] to integrate the diffraction data. Finer sampling of the reciprocal space was used instead with a tilt step of 0.5°. Crystals were fished from the acetone suspension on a lacey-carbon Cu transmission electron microscopy (TEM) grid. Measurements were done in a micro-diffraction mode. Datasets were processed with the PETS program [27], indexed and refined in Jana2006 software package [28].

High-resolution scanning electron microscopy (SEM) was performed using a FEI Nova NanoSEM (Thermo Fisher Scientific, Waltham, MA, USA) equipped with a circular backscatter detector in the backscattered electron mode. An accelerating voltage was set to 5 kV. The samples were suspended in acetone in an ultrasonic bath, deposited onto a silicon wafer chip, and air-dried overnight. Thermal analyses (DTA/TGA) were carried out on a Setaram SETSYS Evolution-16-MS (Setaram, Caluire, France) instrument coupled with a mass spectrometer. The measurements were performed in synthetic air (30 mL min^{-1}) from 20 to 650 °C with a heating rate of 10 °C min^{-1}. Fourier transform infrared (FTIR) spectra were collected with a Nicolet NEXUS 670-FT spectrometer (Thermo Fisher Scientific, Waltham, MA, USA) in KBr pellets. Elemental analysis (CHN) was done using analyzer Thermo Scientific FlashSmartTM 2000 Elemental analyzer (Thermo Fisher Scientific, Waltham, MA, USA). A combustion tube was equipped by EA-2000 chromium oxidizer, high quality copper reducer, and silver cobaltous–cobaltic oxide. All the measurements were performed three times.

The polycrystalline bulk samples were measured at room temperature on a PANalytical Empyrean powder diffractometer (Malvern Pananalytical, Almelo, Netherlands) equipped with a Cu anode and a PIXCel3D detector (Malvern Pananalytical, Almelo, Netherlands). The sample was measured in the reflection Bragg–Brentano geometry. The result of the EDT experiment was confirmed by a simple Rietveld fit in Jana2006, where the structural model was fixed and only profile parameters together with unit cell and preferred orientation were refined, see Figure S1 with the final Rietveld plots. The solid state NMR (ssNMR) spectra were recorded at 11.7 T using a Bruker AVANCE III HD spectrometer (Bruker, Billerica, MA, USA). The 3.2-mm cross-polarization magic angle spinning (CP/MAS) probe was used for ^{13}C and ^{31}P characterization at Larmor frequencies of v (^{13}C) = 125.783 MHz and v (^{31}P) = 202.478 MHz, respectively. The ^{13}C and ^{31}P ssNMR spectra were collected at 20 kHz spinning speed. The ^{13}C NMR isotropic chemical shifts were calibrated using α-glycine (^{13}C: 176.03 ppm; carbonyl signal) and ^{31}P NMR shift using liquid 85% H_3PO_4 in H_2O at 0.0 ppm, both as external standards. High-power ^1H decoupling rCw^{ApA} [29] and SPINAL64 were used to eliminate heteronuclear dipolar couplings, for ^{13}C and ^{31}P ssNMR spectra, respectively. Directly excited ^{13}C and ^{31}P MAS NMR experiments were performed with a recycle delay of 10 s. Five thousand one hundred and twenty and 128 scans were collected in each ^{13}C and ^{31}P MAS NMR experiment, respectively. The total of 4096 scans were accumulated for each ^{13}C CP/MAS NMR spectrum with a recycle delay of 2 s and 1.75 ms cross-polarization (CP) mixing time. Each ^{31}P CP/MAS NMR spectrum was recorded using 512 scans with a recycle delay of 2 s and 2 ms CP mixing time. Frictional heating of the spinning samples was compensated by active cooling [30] and the dried samples were placed into ZrO_2 rotors. All NMR spectra were processed using the Top Spin 3.5 pl7 software package.

3. Results and Discussion

3.1. Synthesis

The coordination polymer ICR-9 with the formula $Ce_2(C_8H_{10}P_2O_4)_3$ was prepared by the solvothermal synthesis. $H_2PBP(Me)$ was dissolved in H_2O and the formed solution was preheated to 100 °C, the DMF solution of Ce(IV) salt was added, and the reaction mixture was stirred at 100 °C for 3 h. The produced sample is further denoted as ICR-9Cryst. We used $(NH_4)_2Ce(NO_3)_6$ for the standard synthesis where Ce(IV) is reduced in situ to Ce(III), similarly as described earlier [31]. To confirm this hypothesis, we performed the syntheses using $Ce(NO_3)_3·6H_2O$ or $CeCl_3·7H_2O$, also yielding materials with the ICR-9 structure. Interestingly, when we exchanged the solvents in which the components were dissolved, i.e., $H_2PBP(Me)$ was dissolved in DMF, preheated, and then Ce(IV) salt was added in H_2O, we obtained the mixture of the UMOF and ICR-9 phases. The amount of the crystalline ICR-9 increased with increasing reaction time. ICR-9A and ICR-9B were prepared with reaction times of 30 and 90 min, respectively. When the reaction was prolonged to 3 h, diffraction lines of additional unknown phase appeared (Figure S2).

3.2. Structure of ICR-9

The structure was solved ab initio in the $P6_3/m$ spacegroup using the SIR2014 software [32]. The lattice parameters were a = 17.4 (1), c = 40.7 (5) Å. Three out of five independent linker molecules were found in the solutions completely. The phenylene groups of the two remaining linkers were not clearly observed in the solution, the atomic positions were found in the difference potential map of the kinematically-refined structure model. The molecules were restrained using distance (0.001 sigma), angular (0.01 sigma), atomic displacement parameters (ADP), and planarity parameters. Hydrogen atoms were introduced in geometrically expected positions and their ADPs were set as riding with extension equal to two. Kinematical refinement resulted in $R_{(obs)}$ = 33.17% on 1704 observed reflections out of 3349 measured using 150 parameters. For crystallographic details see Table S1, the crystallographic data for ICR-9 are deposited at the Cambridge Crystallographic Data Centre (CCDC) no. 1913129 [33].

The crystal structure of ICR-9 consists of 2D layers of octahedrally coordinated cerium atoms bound together through O-P-O bridges, see Figure 1 right. The octahedral coordination of cerium atom is noticeably distorted. The phenylene bridges connect these 2D layers of cerium atoms into the 3D framework (Figure 1 left). There are two rows of differently oriented phenylene bridges. In the first row, containing a mirror plane perpendicular to [001], all three crystallographically independent phenylene groups are arranged perpendicularly to this mirror plane and thus to the 2D layers of cerium atoms, while in the second row the angles between bridging phenylene groups and 2D connecting layers are 67° and 75°. The structure contains voids with a calculated pore limiting diameter of approximately 1.3 Å and a maximal pore diameter in cavities of approximately 3.6 Å. Although the calculated values may be affected by an error, they confirm the nonporous nature of IRC-9Cryst (see below), because the size of the pore limiting diameter is too small for N_2 molecules.

The structure of ICR-9 has a similar 2D layered arrangement to Ce(III) coordinated by 1,4-phenylenebis(phosphonate) with the formula $Ce[O_3P(C_6H_4)PO_3H]$, see Figure 2 [34]. In this case, Ce(III) ions are coordinated by 8 oxygen atoms and create a slightly distorted dodecahedron bisdisphenoid. The connectivities in these layers are different than those in ICR-9. The dodecahedron bisdisphenoids are connected directly with each other through edge-sharing on two opposite sides and by O-P-O bridges forming the final 2D layer. The layers are connected by phenylene groups that are disordered in two positions. To the best of our knowledge, ICR-9 is the first structure with the motive of 2D layers of cerium atoms arranged in the hexagonal honeycomb, connected by oxygen atoms through phosphorus or through any other atoms —(O—Ce—O—X)$_6$— where X = P, S, C, or any other atom.

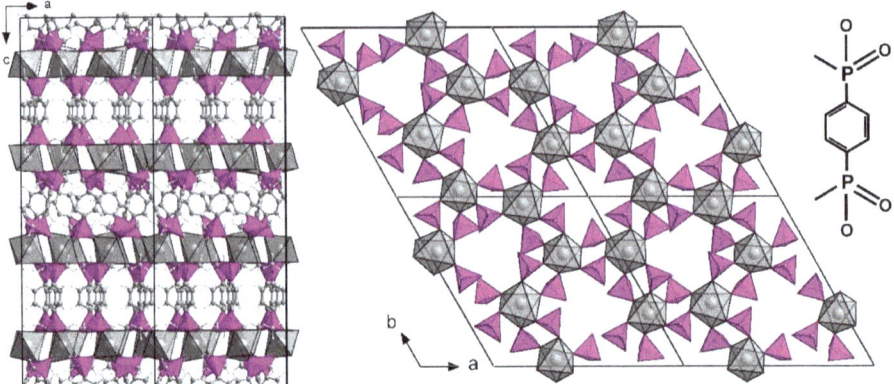

Figure 1. Cluster of 2 × 2 × 1 unit cells of the 2D layer of cerium atoms arranged in the hexagonal honeycomb. The cerium atoms are octahedrally coordinated with oxygen atoms (grey polyhedra) and bound together through O-P-O bridges with phosphorus in a tetrahedral environment (magenta polyhedra). Left is the view along the b direction. Two types of PBP(Me)$^{2-}$ bridging rows are clearly visible: (i) phenylene groups are perpendicular to the 2D layers of cerium atoms, (ii) the angles between phenylene groups and the connecting 2D layer are approximately 70°. Right is the view along the c direction, carbon atoms of phenylene groups and hydrogen atoms were removed for better clarity. The schematic representation of the linker is in the top right corner.

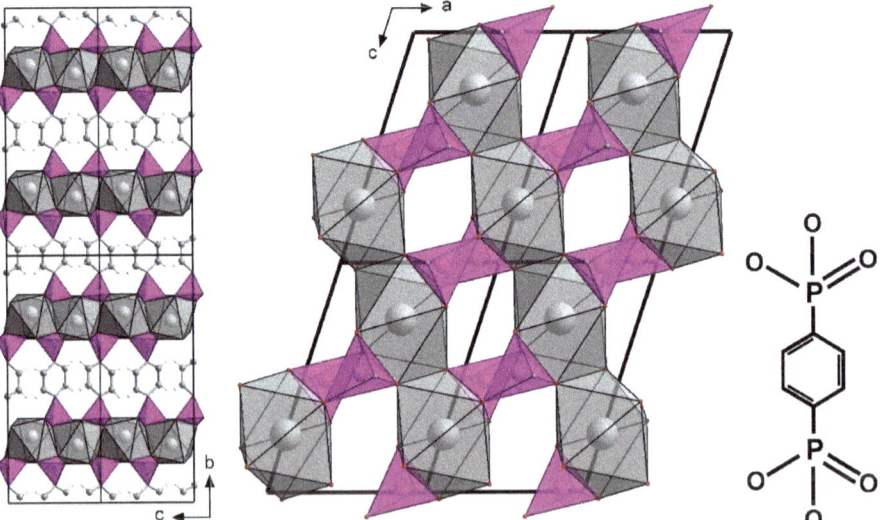

Figure 2. Crystal structure of 1,4-phenylenebis(phosphonate) Ce(III) with the Cambridge Structural Database (CSD) reference code MAXFEY consisting of 2D layers that are connected by the linker [34]. From the side view along the a direction, the crystal structure is similar to ICR-9. However, the coordination of Ce(III) (grey polyhedra) by oxygen atoms as well as its arrangement in the 2D layer are different in comparison with octahedral coordination and honeycomb arrangement of Ce(III) in ICR-9. The tetrahedral environment of phosphorus atoms is depicted by magenta polyhedra. The schematic representation of the linker is in the bottom right corner.

3.3. Characterization

The purity of the bulk ICR-9Cryst sample was confirmed by a simple Rietveld fit using fundamental parameters approach, see Figure S1. Interestingly, the comparison of the full width at half maximum of Bragg reflections of ICR-9Cryst, ICR-9A, and ICR-9B revealed, that the crystalline domain size does not change significantly (approximately 170 nm). The most obvious difference is the presence of a broad peak of amorphous phase between 6 and 12° 2θ in the ICR-9A and ICR-9B patterns. The peak / background ratio indicate various amounts of an amorphous phase, see Figure 3 [35,36]. While ICR-9Cryst does not contain the broad amorphous peak, the largest amount of the amorphous phase is present in the ICR-9A sample, see Table 1.

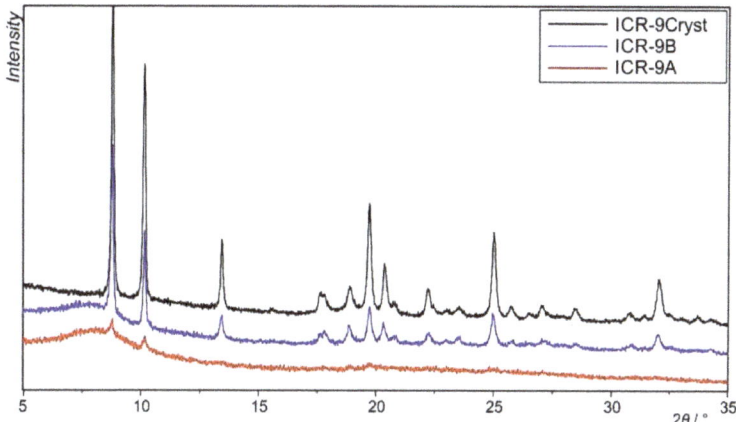

Figure 3. Comparison of the powder XRD patterns of ICR-9Cryst (black), ICR-9B (blue), and ICR-9A (red). The amount of the amorphous phase can be distinguished from the peak / background ratio and the wide peak between 6 and 12° 2θ. The least amount of the amorphous phase is in the sample of ICR-9Cryst, whereas the largest amount is in the case of ICR-9A.

Table 1. Physicochemical properties of ICR-9Cryst, ICR-9A, and ICR-9B.

Sample	mol% DMF[a]*	mol% of Uncoordinated PBP(Me)$^{2-}$ [b]*	Specific Surface Area / m^2 g^{-1}	Estimation of the Amorphous Content / wt%[c]**
ICR-9Cryst	<5%	1%	13	0%
ICR-9B	37%	21%	211	36%
ICR-9A	31%	61%	395	77%

[a] The mol% of DMF were determined from Equation (1). [b] The mol% of uncoordinated PBP(Me)$^{2-}$ were determined from integral areas of individual peaks of ^{31}P MAS NMR spectra. [c] Estimation of the amorphous content was calculated as the intensity ratio of the diffraction Bragg peaks (I_{net}) and of the sum of all measured intensity (I_{tot}) from which the constant background intensity (I_{bg}) was subtracted [1 - $I_{net}/(I_{tot}-I_{bg})$] * Determined experimental error of mol% was estimated to ± 5% in both cases. ** Precision of this method strongly depends on the determination of the constant background intensity. The background from the Rietveld fitting of ICR-9Cryst was used for that purpose.

As mentioned above, the modification of the synthetic conditions led to the microporosity of ICR-9A and ICR-9B. The specific surface area ranged from 13 m^2 g^{-1} for the well crystalline ICR-9Cryst up to 395 m^2 g^{-1} for ICR-9A (Figure 4). The pore size distributions of ICR-9A and ICR-9B are rather wide with a maximum at 7 Å, see Figures S3 and S4.

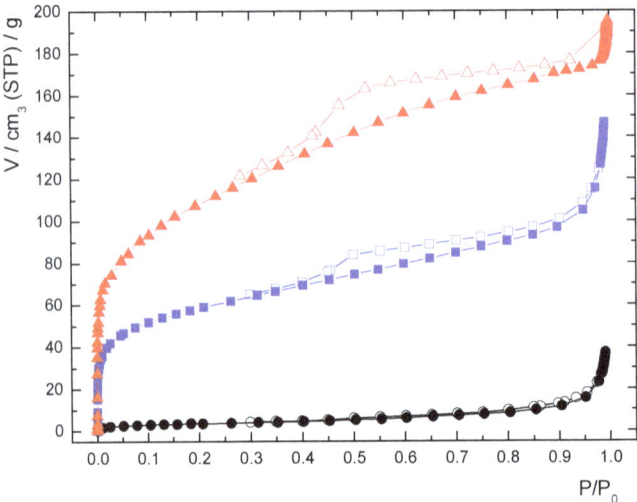

Figure 4. Nitrogen adsorption isotherms for ICR-9Cryst (black), ICR-9B (blue), and ICR-9A (red) at 77 K; adsorption is marked with full symbols and desorption with empty symbols.

The SEM image of ICR-9Cryst (Figure 5) shows uniform, well-shaped particles with hexagonal structure. In the case of ICR-9A, the uniformity of the particles is lost, and sheet-like particles are accompanied by a phase of an unresolved shape. Similarly, ICR-9B formed sheet-like particles. However, the presence of the phase of unresolved shape is limited.

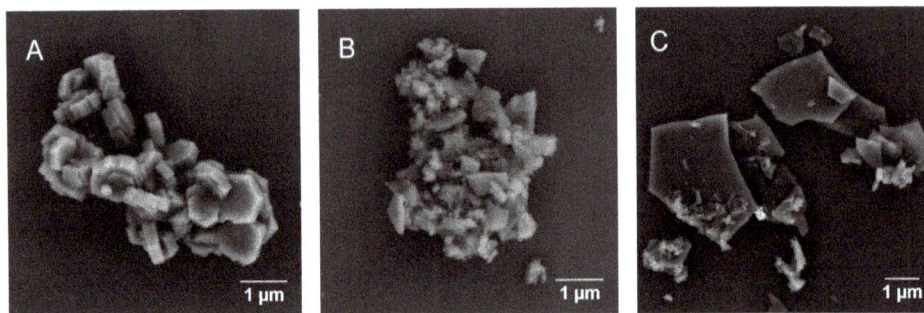

Figure 5. SEM images of ICR-9Cryst (A), ICR-9A (B), and ICR-9B (C).

To describe the UMOF phase and ascertain the origin of the microporosity, we performed detailed solid state NMR (ssNMR) study. The presented ^{13}C and ^{31}P ssNMR (directly-excited MAS and CP/MAS NMR) spectra indicate the formation of a 3D-coordination polymer with a different level of defects, in other words with distinct crystallinity and porosity. In Figure 6, ^{13}C ssNMR spectra (a) and ^{31}P ssNMR spectra (b) of all prepared ICR-9 samples are depicted. The ^{13}C MAS and CP/MAS NMR spectra (Figure 6a,c) confirm the presence of PBP(Me)$^{2-}$ incorporated into the structure in all cases. The signals at 142 ± 2 ppm and 130 ± 1 ppm in all the ^{13}C ssNMR spectra were attributed to individual non-equivalent aromatic carbons, (>P-C$_{Ar}$≤ and = CH$_{Ar}$-), respectively. Furthermore, in the case of the highly-crystalline system (ICR-9Cryst), three distinct methyl groups are clearly resolved with chemical shifts 28.5 ppm, 20.2 ppm, and 16.4 ppm. The presence of these methyl groups can be expected in all samples, and this is verified based on the shape of the unresolved signal in the relevant region (Figure 6c). Interestingly, when a cross polarization technique (^{13}C CP/MAS NMR) was used the resonance at 28.5 ppm disappeared, which indicates higher mobility of these methyl groups in comparison with the other two. Moreover, in the cases of samples containing the UMOF phase (ICR-9A and ICR-9B), a considerable amount of DMF was also detected in the ^{13}C MAS NMR spectra. The presence of DMF was further confirmed by DTA/TGA (Figures S5–S7) and elemental analyses. The amount of DMF in the ICR systems was determined using Equation 1 and is listed in Table 1.

$$mol\%(DMF) = \frac{\left(I_{(CHO)}\right)_{DMF}}{\left(\frac{I_{(CH)Ar}}{4}\right)_{PBP(Me)}}, \quad (1)$$

where $(I_{(CHO)})_{DMF}$ corresponds to the integral area of peaks at 167.1 ppm attributed to the DMF aldehyde group. The integral area of the peak at 130 ± 1 ppm is attributed to the four CH groups of the aromatic rings and is marked as $(I_{(CH)Ar}/4)_{PBP(Me)}$.

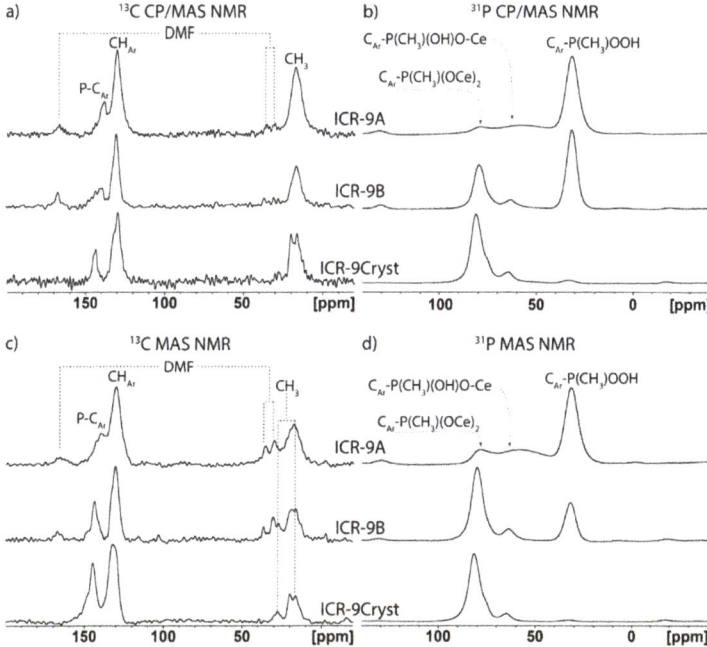

Figure 6. ^{13}C and ^{31}P solid state NMR (ssNMR) spectra of ICR-9Cryst, ICR-9B, and ICR-9A: (a) ^{13}C cross-polarization magic angle spinning (CP/MAS) NMR; (b) ^{31}P CP/MAS NMR; (c) ^{13}C MAS NMR; (d) ^{31}P MAS NMR experiments.

In the ^{31}P MAS and CP/MAS NMR spectra, three distinct signals in a relatively broad range of chemical shifts were recorded (Figure 6b, d). These signals correspond to phosphinate groups of PBP(Me)$^{2-}$ in three different coordination modes: the signals at 80.6 ± 2 ppm, 61.6 ± 4 ppm, and 31.5 ± 1 ppm in the ^{31}P ssNMR spectra were attributed to phosphinate groups coordinated by two oxygens, one oxygen, and non-coordinated phosphinate groups of PBP(Me)$^{2-}$ to Ce(III) atoms, respectively. This assignment of the individual peaks is based on the ^{31}P NMR spectrum of neat PBP(Me)$^{2-}$ and on a significant enhancement of signal intensities at 61.6 ± 4 ppm and 31.5 ± 1 ppm when a cross polarization technique (^{31}P CP/MAS NMR) was employed (Figure S8). The increased intensity indicates the presence of hydroxyl groups in close proximity to phosphorus atoms. Furthermore, the observable change of ^{31}P NMR chemical shifts between "non-coordinated" PBP(Me)$^{2-}$ and neat

PBP(Me)$^{2-}$ as well as the half-width of peaks at 31.5 ± 1 ppm in ^{31}P ssNMR spectra suggests that PBP(Me)$^{2-}$ is coordinated into the framework structure by at least one functional group. Simply put, the presence of totally uncoordinated PBP(Me)$^{2-}$ is excluded in all three investigated systems. On the other hand, the amount of uncoordinated phosphinate groups corresponds to the increasing amount of the UMOF phase as well as to the increasing porosity (Table 1).

4. Conclusions

In the present work, we have prepared the first cerium phosphinate coordination polymer ICR-9. For this purpose, we used phenylene-1,4-bis(methylphosphinic acid) to obtain the polymer with the Ce$_2$[PBP(Me)]$_3$ formula. The structure was determined by electron diffraction tomography to reveal an unusual coordination motive of 2D layers with octahedrally coordinated cerium atoms arranged in the hexagonal honeycomb array. These layers are connected by phenylene bridges to form the 3D polymer. The structure is rather dense with pores smaller than the size of N$_2$ molecules; however, when the unconventional MOF is formed from the same components, the specific surface area can be as high as approximately 400 m^2 g^{-1}.

Supplementary Materials: The following are available online at http://www.mdpi.com/2073-4352/9/6/303/s1, Figure S1: Rietveld fit of ICR-9Cryst. Figure S2: Powder XRD pattern of the sample prepared by the defective procedure with a reaction time of 3 h. Figure S3: Pore size distribution of ICR-9A. Figure S4: Pore size distribution of ICR-9B. Figure S5: DTA/TGA curves and the evolution of gases for ICR-9Cryst. Figure S6: DTA/TGA curves and the evolution of gases for ICR-9A. Figure S7: DTA/TGA curves and the evolution of gases for ICR-9B. Figure S8: Assignment of ^{31}P peaks in the solid state NMR spectra. Figure S9: FTIR spectra of ICR-9Cryst, ICR-9A, and ICR-9B. Table S1. Crystallographic details of ICR-9.

Author Contributions: Conceptualization, J.D.; investigation, J.R., D.B., L.K., P.B, and J.H.; writing—original draft preparation, J.D. and J.R.; writing—review and editing, K.L. and J.B.; project administration, J.D.

Funding: This research was funded by the Czech Science Foundation, grant number 18-12925S.

Acknowledgments: The authors are grateful to the working group Interactions of Inorganic Clusters, Cages, and Containers with Light within the AV21 Strategy of the Czech Academy of Science, Petr Bezdička for the measurement of powder XRD, and the use of ASTRA laboratory instruments established within the Operation program Prague Competitiveness - project CZ.2.16/3.1.00/24510.

Conflicts of Interest: The authors declare no conflict of interest.

References

1. Kaskel, S. *The Chemistry of Metal–Organic Frameworks: Synthesis, Characterization, and Applications*; Wiley-VCH: Weinheim, Germany, 2017; Volume 2.
2. Leus, K.; Bogaerts, T.; De Decker, J.; Depauw, H.; Hendrickx, K.; Vrielinck, H.; Van Speybroeck, V.; Van Der Voort, P. Systematic study of the chemical and hydrothermal stability of selected "stable" Metal Organic Frameworks. *Microporous Mesoporous Mater.* **2016**, *226*, 110–116. [CrossRef]
3. Bůžek, D.; Demel, J.; Lang, K. Zirconium Metal-Organic Framework UiO-66: Stability in Aqueous Environment and its Relevance for Organophosphate Degradations. *Inorg. Chem.* **2018**, *57*, 14290–14297. [CrossRef] [PubMed]
4. Bůžek, D.; Zelenka, J.; Ulbrich, P.; Ruml, T.; Křížová, I.; Lang, J.; Kubát, P.; Demel, J.; Kirakci, K.; Lang, K. Nanoscaled porphyrinic metal-organic frameworks: Photosensitizer delivery systems for photodynamic therapy. *J. Mater. Chem. B* **2017**, *5*, 1815–1821. [CrossRef]
5. Clearfield, A.; Demadis, K. *Metal Phosphonate Chemistry: From Synthesis to Applications*; Royal Society of Chemistry: Oxford, UK, 2012.
6. Taddei, M.; Costantino, F.; Vivani, R. Robust Metal-Organic Frameworks Based on Tritopic Phosphonoaromatic Ligands. *Eur. J. Inorg. Chem.* **2016**, *2016*, 4300–4309. [CrossRef]
7. Shearan, S.J.I.; Stock, S.; Emmerling, F.; Demel, J.; Wright, P.A.; Demadis, K.D.; Vassaki, M.; Costantino, F.; Vivani, R.; Sallard, S.; et al. New Directions in Metal Phosphonate and Phosphinate Chemistry. *Crystals* **2019**, *9*, 270. [CrossRef]
8. Gagnon, K.J.; Perry, H.P.; Clearfield, A. Conventional and Unconventional Metal-Organic Frameworks Based on Phosphonate Ligands: MOFs and UMOFs. *Chem. Rev.* **2012**, *112*, 1034–1054. [CrossRef] [PubMed]

9. Carson, I.; Healy, M.R.; Doidge, E.D.; Love, J.B.; Morrison, C.A.; Tasker, P.A. Metal-binding motifs of alkyl and aryl phosphinates; versatile mono and polynucleating ligands. *Coord. Chem. Rev.* **2017**, *335*, 150–171. [CrossRef]
10. Du, J.-L.; Rettig, S.J.; Thompson, R.C.; Trotter, J. Synthesis, structure, and magnetic properties of diphenylphosphinates of cobalt(II) and manganese(II). The crystal and molecular structures of the γ forms of poly-bis(μ-diphenylphosphinato)cobalt(II) and manganese(II). *Can. J. Chem.* **1991**, *69*, 277–285. [CrossRef]
11. Rosca, I.; Nechita, M.-T.; Sutiman, D.; Cailean, A.; Sibiescu, D.; Vizitiu, M. New iron (III) coordination compounds with applications in water treatment. *Environ. Eng. Manag. J.* **2010**, *9*, 511–517.
12. Oleshkevich, E.; Viñas, C.; Romero, I.; Choquesillo-Lazarte, D.; Haukka, M.; Teixidor, F. M-Carboranylphosphinate as Versatile Building Blocks to Design all Inorganic Coordination Polymers. *Inorg. Chem.* **2017**, *56*, 5502–5505. [CrossRef]
13. Oleshkevich, E.; Teixidor, F.; Rosell, A.; Viñas, C. Merging Icosahedral Boron Clusters and Magnetic Nanoparticles: Aiming toward Multifunctional Nanohybrid Materials. *Inorg. Chem.* **2018**, *57*, 462–470. [CrossRef] [PubMed]
14. Du, Z.-Y.; Zhang, L.; Wang, B.-Y.; Liu, S.-J.; Huang, B.; Liu, C.-M.; Zhang, W.-X. Two magnetic Δ-chain-based Mn(II) and Co(II) coordination polymers with mixed carboxylate-phosphinate and μ_3-OH$^-$ bridges. *CrystEngComm.* **2017**, *19*, 1052–1057. [CrossRef]
15. Yang, W.; Wang, H.; Tian, W.-G.; Li, J.; Sun, Z.-M. The first family of actinide carboxyphosphinates: Two- and three-dimensional uranyl coordination polymers. *Eur. J. Inorg. Chem.* **2014**, *31*, 5378–5384. [CrossRef]
16. Li, J.; Xue, C.-C.; Liu, S.; Wang, Z.-X. Structures and magnetic properties of two noncentrosymmetric coordination polymers based on carboxyphosphinate ligand. *Solid State Sci.* **2016**, *61*, 111–115. [CrossRef]
17. Cecconi, F.; Dakternieks, D.; Duthie, A.; Ghilardi, C.A.; Gili, P.; Lorenzo-Luis, P.A.; Midollini, S.; Orlandini, A. Inorganic-organic hybrids of the p,p'-diphenylmethylenediphosphinate ligand with bivalent metals: A new 2D-layered phenylphosphinate zinc(II) complex. *J. Solid State Chem.* **2004**, *177*, 786–792. [CrossRef]
18. Shekurov, R.; Miluykov, V.; Kataeva, O.; Krivolapov, D.; Sinyashin, O.; Gerasimova, T.; Katsyuba, S.; Kovalenko, V.; Krupskaya, Y.; Kataev, V.; et al. Inorganic-organic hybrids of the p,p'-diphenylmethylenediphosphinate ligand with bivalent metals: A new 2D-layered phenylphosphinate zinc(II) complex. *Cryst. Growth Des.* **2016**, *16*, 5084–5090. [CrossRef]
19. Shekurov, R.; Khrizanforova, V.; Gilmanova, L.; Khrizanforov, M.; Miluykov, V.; Kataeva, O.; Yamaleeva, Z.; Burganov, T.; Gerasimova, T.; Khamatgalimov, A.; et al. Zn and Co redox active coordination polymers as efficient electrocatalysts. *Dalton Trans.* **2019**, *48*, 3601–3609. [CrossRef] [PubMed]
20. Hynek, J.; Brázda, P.; Rohlíček, J.; Londesborough, M.G.S.; Demel, J. Phosphinic Acid Based Linkers: Building Blocks in Metal–Organic Framework Chemistry. *Angew. Chem. Int. Ed.* **2018**, *130*, 5016–5019. [CrossRef]
21. Costantino, F.; Gentili, P.L.; Audebrand, N. A new dual luminescent pillared cerium(IV)sulfate–diphosphonate. *Inorg. Chem. Commun.* **2009**, *12*, 406–408. [CrossRef]
22. Lammert, M.; Wharmby, M.T.; Smolders, S.; Bueken, B.; Lieb, A.; Lomachenko, K.A.; De Vos, D.; Stock, N. Cerium-based metal organic frameworks with UiO-66 architecture: Synthesis, properties and redox catalytic activity. *Chem. Commun.* **2015**, *51*, 12578–12581. [CrossRef]
23. Lammert, M.; Reinsch, H.; Murray, C.A.; Wharmby, M.T.; Terraschke, H.; Stock, N. Synthesis and Structure of Zr(IV)- and Ce(IV)-Based CAU-24 with 1,2,4,5-Tetrakis(4-Carboxyphenyl)-Benzene. *Dalton Trans.* **2016**, *45*, 18822–18826. [CrossRef] [PubMed]
24. Almáši, M.; Zeleňák, V.; Opanasenko, M.; Císařová, I. Ce(III) and Lu(III) Metal-Organic Frameworks with Lewis Acid Metal Sites: Preparation, Sorption Properties and Catalytic Activity in Knoevenagel Condensation. *Catal. Today* **2015**, *243*, 3098–3114. [CrossRef]
25. Atzori, C.; Lomachenko, K.A.; Øien-Ødegaard, S.; Lamberti, C.; Stock, N.; Barolo, C.; Bonino, F. Disclosing the Properties of a New Ce(III)-Based MOF: Ce$_2$(NDC)$_3$(DMF)$_2$ *Cryst. Growth Des.* **2019**, *19*, 787–796. [CrossRef]
26. Vincent, R.; Midgley, P.A. Double conical beam-rocking system for measurement of integrated electron diffraction intensities. *Ultramicroscopy* **1994**, *53*, 271–282. [CrossRef]
27. Palatinus, L. *PETS–Program for Analysis of Electron Diffraction Data*; Institute of Physics of the Czech Academy of Sciences: Prague, Czech Republic, 2011.
28. Petříček, V.; Dušek, M.; Palatinus, L. Crystallographic Computing System JANA2006: General features. *Z. Kristallogr.* **2014**, *229*, 345–352. [CrossRef]

29. Equbal, C.A.; Bjerring, M.; Madhu, P.K.; Nielsen, N.C. Improving spectral resolution in biological solid-state NMR using phase-alternated rCW heteronuclear decoupling. *Chem. Phys. Lett.* **2015**, *635*, 339–344. [CrossRef]
30. Brus, J. Heating of Samples induced by fast magic-angle spinning. *Solid State Nucl. Magn. Reson.* **2000**, *16*, 151–160. [CrossRef]
31. Rhauderwiek, T.; Heidenreich, N.; Reinsch, H.; Øien-Ødegaard, S.; Lomachenko, K.A.; Rütt, U.; Soldatov, A.V.; Lillerud, K.P.; Stock, N. Co-Ligand Dependent Formation and Phase Transformation of Four Porphyrin-Based Cerium Metal–Organic Frameworks. *Cryst. Growth Des.* **2017**, *17*, 3462–3474. [CrossRef]
32. Burla, M.C.; Caliandro, R.; Carrozzini, B.; Cascarano, G.L.; Cuocci, C.; Giacovazzo, C.; Mallamo, M.; Mazzone, A.; Polidori, G. Crystal structure determination and refinement via SIR2014. *J. Appl. Cryst.* **2015**, *48*, 306–309. [CrossRef]
33. CCDC no. 1913129. Available online: http://www.ccdc.cam.ac.uk/conts/retrieving.html (accessed on 30 April 2019).
34. Amghouz, Z.; García-Granda, S.; García, J.R.; Clearfield, A.; Valiente, R. Organic-Inorganic Hybrids Assembled from Lanthanide and 1,4-Phenylenebis(phosphonate). *Cryst. Growth Des.* **2011**, *11*, 5289–5297. [CrossRef]
35. Bennett, T.D.; Cheetham, A.K. Amorphous Metal–Organic Frameworks. *Acc. Chem. Res.* **2014**, *47*, 1555–1562. [CrossRef] [PubMed]
36. Shearer, G.C.; Chavan, S.; Bordiga, S.; Svelle, S.; Olsbye, U.; Lillerud, K.P. Defect Engineering: Tuning the Porosity and Composition of the Metal–Organic Framework UiO-66 via Modulated Synthesis. *Chem. Mater.* **2016**, *28*, 3749–3761. [CrossRef]

© 2019 by the authors. Licensee MDPI, Basel, Switzerland. This article is an open access article distributed under the terms and conditions of the Creative Commons Attribution (CC BY) license (http://creativecommons.org/licenses/by/4.0/).

MDPI
St. Alban-Anlage 66
4052 Basel
Switzerland
Tel. +41 61 683 77 34
Fax +41 61 302 89 18
www.mdpi.com

Crystals Editorial Office
E-mail: crystals@mdpi.com
www.mdpi.com/journal/crystals

www.ingramcontent.com/pod-product-compliance
Lightning Source LLC
LaVergne TN
LVHW071959080526
838202LV00064B/6789